Lecture Notes in Artificial Intelligence 10097

Subseries of Lecture Notes in Computer Science

LNAI Series Editors

Randy Goebel
University of Alberta, Edmonton, Canada
Yuzuru Tanaka
Hokkaido University, Sapporo, Japan
Wolfgang Wahlster
DFKI and Saarland University, Saarbrücken, Germany

LNAI Founding Series Editor

Joerg Siekmann
DFKI and Saarland University, Saarbrücken, Germany

More information about this series at http://www.springer.com/series/1244

Wei Lee Woon · Zeyar Aung
Oliver Kramer · Stuart Madnick (Eds.)

Data Analytics
for Renewable
Energy Integration

4th ECML PKDD Workshop, DARE 2016
Riva del Garda, Italy, September 23, 2016
Revised Selected Papers

 Springer

Editors
Wei Lee Woon
Masdar Institute of Science and Technology
Abu Dhabi
United Arab Emirates

Zeyar Aung
Electrical Engineering and Computer
   Science
Masdar Institute of Science and Technology
Abu Dhabi
United Arab Emirates

Oliver Kramer
Informatik
Universität Oldenburg
Oldenburg
Germany

Stuart Madnick
Massachusetts Institute of Technology
Cambridge, MA
USA

ISSN 0302-9743   ISSN 1611-3349 (electronic)
Lecture Notes in Artificial Intelligence
ISBN 978-3-319-50946-4   ISBN 978-3-319-50947-1   (eBook)
DOI 10.1007/978-3-319-50947-1

Library of Congress Control Number: 2017930213

LNCS Sublibrary: SL7 – Artificial Intelligence

Printed on acid-free paper

This Springer imprint is published by Springer Nature
The registered company is Springer International Publishing AG
The registered company address is: Gewerbestrasse 11, 6330 Cham, Switzerland

Preface

This volume presents a collection of papers focused on the use of data analytics and machine learning techniques to facilitate the integration of renewable energy resources into existing infrastructure and socioeconomic systems. This volume includes papers presented at DARE 2016, the 4th International Workshop on Data Analytics for Renewable Energy Integration, which was hosted by ECML PKDD 2016, and a few invited articles.

In recent times, climate change, energy security, and sustainability have focused much attention on the development of clean and renewable energy sources. However, of equal importance is the issue of integrating these sources into existing infrastructure and socioeconomic systems. While increasing the generating capacities of renewable energy sources is still important, issues such as efficient and cost-effective storage and distribution, demand response, planning, and policy making must be resolved in parallel. These challenges are inherently multidisciplinary and depend heavily on robust and scalable computing techniques and the ability to handle large, complex data sets. The domains of data analytics, pattern recognition, and machine learning are uniquely positioned to offer solutions to many of these challenges. Examples of relevant topics include time series forecasting, the detection of faults, cyber security, smart grid and smart cities, technology integration, demand response, and many others.

This year's event attracted numerous researchers working in the various related domains, both to present and discuss their findings and to share their respective experiences and concerns. We are very grateful to the organizers of ECML PKDD 2016 for hosting DARE 2016, the Program Committee members for their time and assistance, and the Masdar Institute, MIT, and the University of Oldenburg for their support of this timely and important workshop. Last but not least we sincerely thank the authors for their valuable contributions to this volume.

October 2016

Wei Lee Woon
Zeyar Aung
Oliver Kramer
Stuart Madnick

Organization

Program Chairs

Wei Lee Woon   Masdar Institute of Science and Technology, UAE
Zeyar Aung   Masdar Institute of Science and Technology, UAE
Stuart Madnick   Massachusetts Institute of Technology, USA

Organizers

Wei Lee Woon   Masdar Institute of Science and Technology, UAE
Zeyar Aung   Masdar Institute of Science and Technology, UAE
Oliver Kramer   University of Oldenburg, Germany
Stuart Madnick   Massachusetts Institute of Technology, USA

Program Committee

Abel Sanchez   Massachusetts Institute of Technology, USA
Francisco Martínez Álvarez   Pablo de Olavide University, Spain
Fabian Gieseke   Radboud University, The Netherlands
Jimmy Peng   National University of Singapore, Singapore
Davor Svetinovic   Masdar Institute of Science and Technology, UAE
David Lowe   Aston University, UK
Depeng Li   University of Hawaii at Manoa, USA
Srinivas Sampalli   Dalhousie University, Canada
Paul Yoo   Bournemouth University, UK
Erik Casagrande   GE, UAE
Randa Herzallah   Aston University, UK

Contents

Locating Faults in Photovoltaic Systems Data 1
  Alexander Kogler and Patrick Traxler

Forecasting of Smart Meter Time Series Based on Neural Networks 10
  Thierry Zufferey, Andreas Ulbig, Stephan Koch, and Gabriela Hug

Cybersecurity for Smart Cities: A Brief Review 22
  Armin Alibasic, Reem Al Junaibi, Zeyar Aung, Wei Lee Woon,
  and Mohammad Atif Omar

Machine Learning Prediction of Photovoltaic Energy from Satellite Sources . . . 31
  Alejandro Catalina, Alberto Torres-Barrán, and José R. Dorronsoro

Approximate Probabilistic Power Flow 43
  Carlos D. Zuluaga and Mauricio A. Álvarez

Dealing with Uncertainty: An Empirical Study on the Relevance
of Renewable Energy Forecasting Methods 54
  Robert Ulbricht, Anna Thoß, Hilko Donker, Gunter Gräfe,
  and Wolfgang Lehner

Measuring Stakeholders' Perceptions of Cybersecurity for Renewable
Energy Systems 67
  Stuart Madnick, Mohammad S. Jalali, Michael Siegel, Yang Lee,
  Diane Strong, Richard Wang, Wee Horng Ang, Vicki Deng,
  and Dinsha Mistree

Selection of Numerical Weather Forecast Features for PV Power
Predictions with Random Forests 78
  Björn Wolff, Oliver Kramer, and Detlev Heinemann

Evolutionary Multi-objective Ensembles for Wind Power Prediction 92
  Justin Heinermann, Jörg Lässig, and Oliver Kramer

A Semi-automatic Approach for Tech Mining and Interactive
Taxonomy Visualization 102
  Ioannis Karakatsanis, Alexandros Tsoupos, and Wei Lee Woon

Decomposition of Aggregate Electricity Demand into the Seasonal-Thermal
Components for Demand-Side Management Applications in "Smart Grids" . . . 116
  Andreas Paisios and Sasa Djokic

Author Index 137

Locating Faults in Photovoltaic Systems Data

Alexander Kogler[✉] and Patrick Traxler[✉]

Data Analysis Systems Group, Software Competence Center Hagenberg,
Hagenberg im Mühlkreis, Austria
{alexander.kogler,patrick.traxler}@scch.at

Abstract. Faults of photovoltaic systems often result in an energy drop
and therefore decrease the efficiency of the system. Detecting and analyzing faults is thus an important problem in the analysis of photovoltaic
systems data. We consider the problem of estimating the starting time
and end time of a fault, i.e. we want to locate the fault in time series
data. We assume to know the power output, plane-of-array irradiance
and optionally the module temperature. We demonstrate how to use our
fault location algorithm to classify shading events. We present results on
real data with simulated and real faults.

Keywords: Fault location · Fault detection · Fault diagnosis ·
Photovoltaics · Shading · Sustainable faults · Fault classification ·
Robust regression · Pattern recognition

1 Introduction

Faults affect the performance of photovoltaic (PV) systems since most faults
result in an energy drop, i.e. the PV system converts less solar energy into
electrical energy than possible. Identifying faulty or inefficient PV systems is
thus an important problem in data analysis with applications to maintenance of
PV systems.

In addition to identifying faulty PV systems, we would like to analyze and
explain detected faults. We consider the problem of estimating the starting time
A and end time B of a fault and its application to discovering and classifying
faults. We call this problem *fault location*. Fault location algorithms are of particular interest for analyzing sustainable faults, i.e. faults that occur frequently
over some period of time. An example is shading. Shading usually occurs over
the period of some weeks or months at some particular daytime (e.g. in the afternoon). This pattern allows us to classify a fault as a shading pattern. Classifying
a fault as a shading pattern is an indication for real shading. Such algorithms
are intended to support for example maintenance of PV systems.

The research reported in this paper has been supported by the Austrian Ministry
for Transport, Innovation and Technology, the Federal Ministry of Science, Research
and Economy, and the Province of Upper Austria in the frame of the COMET center
SCCH.

W.L. Woon et al. (Eds.): DARE 2016, LNAI 10097, pp. 1–9, 2017.
DOI: 10.1007/978-3-319-50947-1_1

We restrict to a basic sensor setting. We assume to know the power output P_t, the plane-of-array (POA) irradiance E_t, and optionally the module temperature T_t at time t. We do not make any assumptions on the PV system, i.e. the power inverter, the number of modules or anything else. In addition, we do not assume to have labeled data.

Results. We describe an algorithm for locating faults in PV systems data. We present results for the accuracy and reliability of our algorithm, i.e. how well the algorithm estimates the starting time A and the end time B of some fault. Based on it, we describe a rule-based system for recognizing shading patterns. We present results on real data with real and simulated faults (energy drops). One of our data sets contains verified shading.

A crucial step in our method is robust linear regression. Robustness refers here to a high breakdown point [8]. We compare the effect of robust algorithms on fault location with the effect of non-robust algorithms on fault location. We demonstrate that robust algorithms improve fault location considerably.

Our approach is model- or rule-based. This is due to the lack of labels. Fault classification refers here to recognizing patterns in residuals. We describe these patterns by rules, which requires knowledge about the fault type. Fault detection works similar as in [4,9]. In [9], we detect faults by testing if a PV system behaves as a linear system, i.e. if the system observations fit a linear model. A fault is an energy drop in [4,9] similar as in this work.

Motivation. Our knowledge about operating PV systems is usually very limited. Often we do not know the system design and the power output is the only available sensor information. This is the case for many residential PV systems, but also for solar power facilities. POA-irradiance sensors are also common although clearly not the standard, in particular for residential PV systems. This motivates our restrictive sensor setting. In the open problems section, we discuss the situation that no POA-irradiance sensor is installed.

The motivation for fault classification is that different faults may trigger different actions. Hardware defects such as module failures or sensor defects often require some form of maintenance. Faults such as snow or covering tree leaves are probably not critical and thus do not require any maintenance. Shading is a fault type somewhere in between. A chimney that covers a small fraction of solar modules of a roof-mounted PV system probably reduces the produced energy by a negligible amount of energy. A growing tree however may become or already is a crucial problem for power generation.

1.1   Related Work

Different approaches for detecting faults in PV systems are discussed in several publications like [1–4,9]. However, only a few publications address the problem of fault diagnosis or fault classification. These problems require a more detailed analysis of detected faults.

In order to detect faults (energy drops) of PV systems and to determine their starting time and end time, we use system models similar to those presented in [5]. The authors compare three different models in order to estimate the power output of PV modules.

These models abstract many details of PV systems and only require knowledge of quantities such as power output or POA-irradiance. In other words, it is not necessary to have knowledge of any system parameters of the PV systems.

One of our contributions is the recognition of sustainable, long-lasting faults. Our focus is on shading. Examples for other types of sustainable faults are panel coverage by snow or module defects.

These kinds of faults are also discussed in [4] but the authors pursue an entirely different method. The main difference to our approach is, that a sustainable event is indicated depending on the sun position at fault time. Furthermore, a considerable amount of historical data is required.

The authors of [2] establish a circuit-based PV module which has been developed using simulation software. Among other things, this model is used for generating test data and validating the proposed fault detection approach. This algorithm also determines different fault classes but they only give information about the severity in terms of energy loss.

In [1] fault detection is considered as a clustering task and is tackled by applying the minimum covariance determinant estimator. Like the method proposed in [9] also this approach does not provide for fault diagnosis (classification of faults).

In order to tackle the problem of fault detection and location estimation, we apply different linear regression methods. Regression methods like the Least Trimmed Squares (LTS) estimator [6] and the Least Median of Squares (LMS) estimator [7] are particularly applicable in fault detection because of their robustness against outliers, i.e. they have a high breakdown point [8]. Photovoltaic systems data is frequently contaminated with outliers (faults). In [9] ℓ_2-regression is used for residual analysis and fault detection due to its robustness property.

2 Method and Results

Our method consists of three steps: (1) robust estimation, (2) fault detection and location, and (3) pattern recognition (for shading patterns).

In the first step, we take data of a day as input and generate linear models for every PV system. In the second step, we take data of this day and the linear model computed in step (1) and compute the residuals. We check if during a day a fault happened. If so, we locate faults. Step one and two generate for every day and PV system information whether a fault happened and if so when. In step three, we take these results for up to several weeks and search for patterns in it. In particular, we check if the rules of a shading pattern are satisfied. In what follows, we describe these steps in detail and present experimental results.

2.1   Robust Estimation

In this section we describe PV system models and procedures for estimating unknown parameters. In addition, we present experimental results for the fitness of the models. We consider a discrete daytime setting (5, 10 or 15 min intervals are realistic) and two different linear system models:

$$P_t = a \cdot E_t + b \cdot E_t^2 \tag{1}$$

and

$$P_t = a \cdot E_t + b \cdot E_t^2 + c \cdot T_t + d \cdot E_t \cdot T_t + e \cdot t + f \cdot t^2, \tag{2}$$

where P_t is the power output of the PV system, E_t is the POA-irradiance, T_t is the module temperature, and $t \in I$ is the discrete daytime. For example, if we have measurements for every hour, then $I = \{1, \ldots, 24\}$. The unknown parameters that we want to estimate are a, b, c, d, e, and f.

We consider the Least Trimmed Squares (LTS) estimator [2], which is a robust variant of the Ordinary Least Squares (OLS) estimator. LTS has a breakdown point of 0.5. This means that a bit less than 50% of data points can be arbitrarily corrupted with outliers (faults). Estimations via OLS and LTS are solutions to optimization problems. The objective function of the LTS estimator is the same as for OLS but considers only the 50% smallest squared residuals. We refer to [2] for the exact mathematical definition of LTS.

Table 1 shows experimental results for the LTS estimator and a single PV system. Power output estimates $\hat{P}_t$ are calculated based on energy data of the day that we want to check for potential faults. Model parameters are estimated based on data of the same day. We simulate energy drops by subtracting 10%, 30% and 50% during 2 h of midday from the measured power output. We do this for every day. We calculate the R^2 value a.k.a. coefficient of determination and the model fitness as defined in Eq. 3 for each of 240 days. Table 1 shows the median of these 240 values for each setting.

Besides (ordinary) fitness values we calculate an adapted form of fitness. The adapted fitness is calculated on the 50% smallest squared residuals. It is a robust measure of fitness. We note the adapted R^2-value and fitness in Table 1 since it shows that e.g. the adapted R^2-value remains roughly constant in the presence of an energy drop whereas the R^2-value decreases with an increasing energy drop. This shows that LTS is robust against outliers in the data.

We see in Table 1 that the consideration of module temperature T_t and time t has some effect. In what follows, we thus consider model of Eq. 2.

2.2   Fault Detection and Location

In this section, we treat fault location. The input is data for a PV system and day, i.e. (P_t, E_t, T_t) for $t \in I$, which we defined in Sect. 2.1. Note that t refers to daytime. In addition, we have a system model, parameters a, b, c, d, e and f in Eq. 2, that takes as input the POA-irradiance E_t, module temperature T_t,

Table 1. Model fitness of LTS for a single PV system

	Ordinary fitness				Adapted fitness			
	Model (1)		Model (2)		Model (1)		Model (2)	
	R^2	Fitness	R^2	Fitness	R^2	Fitness	R^2	Fitness
drop = 0%	0.9983	0.9836	0.9989	0.9883	0.9999	0.9947	1	0.9971
drop = 10%	0.9824	0.9548	0.9849	0.9607	0.9999	0.9928	1	0.9961
drop = 30%	0.8578	0.8929	0.8648	0.8997	0.9999	0.9929	1	0.9961
drop = 50%	0.5965	0.8232	0.6096	0.8301	0.9999	0.9928	1	0.9961

time t and outputs an estimate $\hat{P}_t$ for P_t. First, we calculate the residuals R_t as $R_t := P_t - \hat{P}_t$. Next, we detect faults by checking if the *fitness* [9]

$$F := 1 - \frac{\sum_{t \in I} |R_t|}{\sum_{t \in I} |P_t|} \tag{3}$$

is smaller than a threshold θ_{fit}. If $F < \theta_{\text{fit}}$, we say that a fault happened and continue with fault location. Otherwise, we say that no fault happened.

The output of our fault location algorithm Locate, Algorithm 1, are pairs (A_j, B_j), where A_j is the starting time and B_j is the end time of the j-th fault. Algorithm 1 outputs locations for all significant faults. Faults are energy drops. We control the significance by the parameter θ_{sig}, i.e. we output a fault location if the energy lost is significant in relation to the total energy. In our application, it suffices to work only with the three strongest energy drops since in most situations there are at most 3 significant energy drops per day.

The algorithm works by computing *deviation* and continuously checking if $R_i < deviation$ holds to find longest subsequences $I' \subseteq I$ of the residuals such that $R_i < deviation$ for every $i \in I'$. We compute *deviation* on a subset R' of the residuals R due to reasons of robustness.

Figure 1 shows results of the algorithm for a roof-mounted PV system with verified shading. The solid line depicts the measured power output and the dashed line the estimated one. Vertical lines indicate the calculated fault location. A chimney causes the shading that is particularly strong over summer. The energy drops are however relatively small. Detecting and locating such faults is difficult because of the small deviations.

Table 2 shows experimental results for fault location. We consider 240 days and a single PV system. We simulate energy drops, as above, by subtracting 10%, 30% and 50% during 2 h of midday from the measured power output and consider a fault j as correctly located if the estimated starting time A_j and end time B_j differs by at most ± 60 min from the real starting and end time.

We conduct these experiments for another 8 PV systems. Table 3 shows the median of these 8 experiments. We see that the values show a similar behavior as the values in Table 2. Algorithm 1 shows a consistent behavior over all 9 PV systems. The 9 PV systems are all from the same region in Austria.

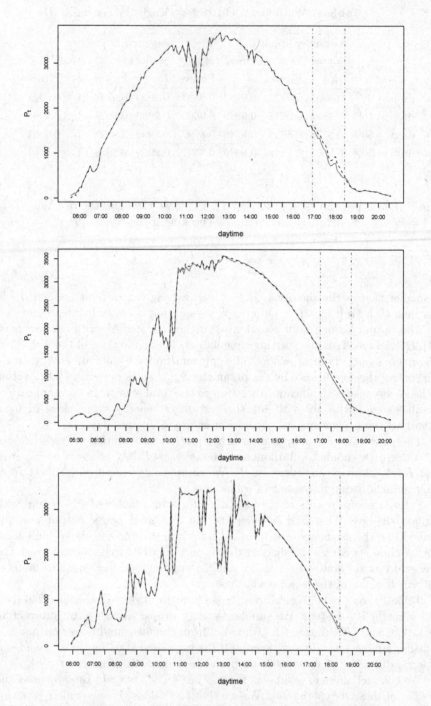

Fig. 1. Three days of a single PV system with verified shading

---

Algorithm 1. Algorithm Locate with parameter θ_{sig} (significance, typically 0.01). Its input is the generated power of the PV system P_i, $i \in I$, for a day and estimates $\hat{P}_i$ for it. Its outputs are the starting and end times of energy drops.

Calculate the total energy $total := \sum_{i \in I} P_i$
Calculate the residuals $R_i := \hat{P}_i - P_i$ for all $i \in I$
Let R' be the elements from R that are the $\lfloor \frac{|I|}{2} \rfloor$-smallest elements in $\{|R_i| : i \in I\}$
Set $deviation := \text{mean}(R') - \text{standard\_deviaton}(R') \cdot 3.0$
Set $j := 1$
for all $i \in I$ **do**
    while $X_i \geq deviation$ **do**
        Increment i
    end while
    $start := i$
    $energy := R_{start}$
    while $R_i < deviation$ **do**
        $energy = energy + R_i$
        Increment i
    end while
    Set $end := i - 1$
    if $energy > \theta_{\text{sig}} \cdot total$ and $i \geq start + 2$ **then**
        $A_j := start$, $B_j := end$, $E_j := energy$
        Increment j
    end if
end for
Output $(A_1, B_1, E_1), (A_2, B_2, E_2) \ldots$

Summarizing the results of Tables 2 and 3, we clearly see that robust linear regression, LTS here, is superior to non-robust linear regression. Furthermore, Algorithm 1 successfully locates most faults, in particular strong faults.

2.3   Pattern Recognition (Shading Patterns)

In this section, we describe a sample application of Algorithm 1. Given that we have detected a fault of a PV system during a day, we want to check if the fault has been possibly caused by shading. The idea is to look for a *shading pattern*. We model the shading pattern by three if-then rules. Let A be the starting time and B the end time of some detected fault.

1. Consider a time period of 7 days before the detected fault. If on at least 4 out of the 7 days some fault occurred with starting time in $A \pm 30$ min and end time in $B \pm 30$ min, then we say that the fault was possibly caused by shading.
2. Consider a time period of 21 days before the detected fault. If on at least 50% of the days in this period, some fault occurred with starting time in $A \pm 1$ h and end time in $B \pm 1$ h, then we say that the fault was possibly caused by shading.

Table 2. Fault location of a single PV system with simulated energy drops and parameters $\theta_{fit} = 0.99$ and $\theta_{sig} = 0.004$

	LTS regression		OLS regression	
	# detected f.	correct loc. (%)	# detected f.	correct loc. (%)
drop = 10%	239	96.23	239	59.41
drop = 30%	239	97.49	239	75.73
drop = 50%	239	98.33	239	78.66

Table 3. Fault location of 8 PV systems with simulated energy drops and parameters $\theta_{fit} = 0.95$ and $\theta_{sig} = 0.004$

	LTS regression		OLS regression	
	# detected f.	correct loc. (%)	# detected f.	correct loc. (%)
drop = 10%	197	48.54	159	12.69
drop = 30%	242	79.34	242	51.03
drop = 50%	244	90.43	244	59.02

3. Consider a time period of 60 days before the detected fault. If at least 50% of detected faults in this time period have starting time in $A \pm 2\,h$, end time $B \pm 2\,h$ and if the number of such events is at least 10, then we say that the fault was possibly caused by shading.

We designed the conditions of the rules in such a way to indicate the confidence of the shading pattern, i.e. if rule (1) applies we are more confident than if rule (2) or (3) applies.

Table 4 shows experimental results concerning the recognition of shading patterns. We applied our method to our 9 PV systems and included the PV systems in the table only if rules fired at least 10 times. PV system (1) contains shading that has been verified on site. The other PV systems with recognized shading patterns have not been verified on site yet, but are likely to be caused by some form of shading. In the last column of Table 4 we noted the number of times at least one of the rules has fired.

Summarizing our results for recognizing shading patterns, we see that our rule-based approach, that is itself based on Algorithm 1, successfully identifies PV systems with shading patterns. In particular, the faults of the PV system with verified shading have been correctly classified as shading patterns.

3   Summary and Open Problems

We presented a method for analyzing PV systems data to locate and classify faults with a focus on shading. We identified fault location as an important step for fault classification. The starting time and end time of a fault are the basis for fault classification that works by identifying predefined patterns, e.g. a pattern indicating real shading. We verified our method on real data.

Table 4. Shading events

	# rule 1	# rule 2	# rule 3	≥ 1 rule
PV system (1)	10	23	33	36
PV system (2)	1	3	23	27
PV system (3)	0	10	30	39
PV system (4)	0	4	27	28
PV system (5)	0	1	25	25

We considered a restrictive sensor setting. We assume to know the power output, the POA-irradiance and optionally the module temperature. An even more restrictive and common sensor setting includes only the power output. Studying this sensor setting remains as an open problem. One possible way to solve this problem is to compare the system behavior to nearby systems [10]. Another way is to integrate irradiance information from satellite images or weather stations. The problem with both ways or a combination of them is the low model accuracy that hinders reliable residual analysis.

We considered the case of recognizing shading patterns that indicate real shading events. Other important faults include covering snow and sustainable hardware defects. It remains as an open problem to classify them.

References

1. Braun, H., Buddha, S.T., Krishnan, V., Spanias, A., Tepedelenlioglu, C., Yeider, T., Takehara, T.: Signal processing for fault detection in photovoltaic arrays. In: 37th IEEE International Conference on Acoustics, Speech and Signal Processing, pp. 1681–1684 (2012)
2. Chao, K.H., Ho, S.H., Wang, M.H.: Modeling and fault diagnosis of a photovoltaic system. Electr. Power Syst. Res. **78**(1), 97–105 (2008)
3. Chouder, A., Silvestre, S.: Fault detection and automatic supervision methodology for PV systems. Energy Convers. Manage. **51**, 1929–1937 (2010)
4. Firth, S., Lomas, K., Rees, S.: A simple model of PV system performance and its use in fault detection. Sol. Energy **84**(4), 624–635 (2010)
5. Marion, B.: Comparison of predictive models for PV module performance. In: 33rd IEEE Photovoltaic Specialist Conference, pp. 1–6 (2008)
6. Mount, D.M., Netanyahu, N.S., Piatko, C.D., Silverman, R., Wu, A.Y.: On the least trimmed squares estimator. Algorithmica **69**(1), 148–183 (2014)
7. Rousseeuw, P.J.: Least median of squares regression. J. Am. Stat. Assoc. **79**(388), 871–880 (1984)
8. Rousseeuw, P.J., Leroy, A.M.: Robust Regression and Outlier Detection. Wiley, Hoboken (2005)
9. Traxler, P.: Fault detection of large amounts of photovoltaic systems. In: Proceedings of the ECML/PKDD 2013 Workshop on Data Analytics for Renewable Energy Integration (2013)
10. Traxler, P., Gomez, P., Grill, T.: A robust alternative to correlation networks for identifying faulty systems. In: Proceedings of the 26th International Workshop on Principles of Diagnosis, pp. 11–18 (2015)

Forecasting of Smart Meter Time Series Based on Neural Networks

Thierry Zufferey[✉], Andreas Ulbig, Stephan Koch, and Gabriela Hug

Power Systems Laboratory, ETH Zurich, Zurich, Switzerland
{thierryz,ulbig,koch,ghug}@eeh.ee.ethz.ch

Abstract. In traditional power networks, Distribution System Operators (DSOs) used to monitor energy flows on a medium- or high-voltage level for an ensemble of consumers and the low-voltage grid was regarded as a black box. However, electric utilities nowadays obtain ever more precise information from single consumers connected to the low- and medium-voltage grid thanks to smart meters (SMs). This allows a previously unattainable degree of detail in state estimation and other grid analysis functionalities such as predictions. This paper focuses on the use of Artificial Neural Networks (ANNs) for accurate short-term load and Photovoltaic (PV) predictions of SM profiles and investigates different spatial aggregation levels. A concluding power flow analysis confirms the benefits of time series prediction to support grid operation. This study is based on the SM data available from more than 40,000 consumers as well as PV systems in the City of Basel, Switzerland.

Keywords: Smart meter · Short-term forecasting · Artificial neural network · Data preparation · Power flow · Big data analytics

1 Introduction

Conventional electricity meters are usually read only once per billing period and give no information as to when energy is consumed at each metered site during that time span. Nevertheless, the current roll-out of new sensor elements called smart meters enables accurate high-resolution measurements on both the spatial scale (on a household level) and the temporal scale (every hour or 15 min) for parts of the distribution grid for which previously only spatially aggregated measurements on the substation and transformer level have been available. At first glance, the main motivations of DSOs to install smart meters are the efficient integration of billing data into the existing billing systems by avoiding manual data gathering, and to facilitate the tracking in case of customers moving to a different property or changing their electricity supplier. Additionally, this evolution can be seen as an excellent opportunity for better operation and planning of active distribution grids, e.g. generation scheduling, load management and system security assessment. This inevitably relies on accurate predictions on the low-voltage level for both distributed generation and end-use consumption, which can be obtained using high-resolution SM data.

© Springer International Publishing AG 2017
W.L. Woon et al. (Eds.): DARE 2016, LNAI 10097, pp. 10–21, 2017.
DOI: 10.1007/978-3-319-50947-1_2

Therefore, this paper presents a comprehensive approach for SM forecasting of several types of loads and PV systems, using big data technologies and parallel cloud computing. Predictions are carried out by means of ANNs for several spatial aggregation sizes, going from individual time series to the sum of all available load or production profiles. Measured time series and forecasting results are finally compared by running a power flow analysis for both cases.

The remainder of this paper is organized as follows: Sect. 2 presents the dataset and indispensable preprocessing tasks before performing the forecasting analysis as described by Sect. 3. This is followed by Sect. 4 which shows a power flow simulation based on predicted time series. Eventually, the main contributions of this paper are summarized and future works are given in Sect. 5.

2   Smart Meter Data Preparation

The data used in this study has been collected by IWB [1], the public utility of the City of Basel, between April 2014 and September 2015 and comes from approximately 40,000 small consumers, 1,000 large consumers (commercial and industrial loads) and 400 PV systems that are well distributed across the city. It is made up of energy consumption values with a sampling period of 15 min. Time series whose data is missing during at least one full day in the case of small consumers or one full month for large consumers and PV systems are discarded. In addition, meteorological data measured each 10 min by MeteoSwiss [2] at the weather station of Binningen is also utilized but has to be adapted to comply with the SM sampling rate.

After the above mentioned removals, it appears that 0.94% of energy values coming from small consumers are missing, which is due to both sporadic connection failures for a few SMs and significant data gaps for a majority of devices during several hours. However, this does not impact the billing process as a separate data register exists for the total yearly energy consumption of a given customer. In order not to introduce a substantial bias into the forecasting process, missing data has to be carefully substituted. Since some SM time series are likely to display similar patterns, weighted K-Nearest Neighbors (KNN) is an appropriate imputation method [3]. For the sake of saving time, a reduced training set of 3,000 normalized time series is first created, from which the 5 closest training examples in terms of Euclidean distance are selected for each incomplete load profile. Missing values are then substituted by the weighted average of the corresponding attribute from the 5 nearest neighbors, i.e. each neighbor contributes proportionally to its proximity degree. The KNN implementation is adapted from the open-source software "Knowledge Extraction based on Evolutionary Learning" (KEEL) [4] and supported by the cloud computing engine Apache Spark [5] deployed on a 16-core Azure Virtual Machine (VM).

An anomaly detection is also carried out. On the one hand, it identifies loads with an unusually low energy consumption, i.e. with an average consumption lower than 100 Wh per day for the dataset with small consumers and lower than 100 kWh per month for the one with large consumers. Since the forecasting algorithm performs very poorly on these load profiles whose energy share among all customers is in fact negligible, they can be excluded from the study. On the other hand, large consumers with a share of zero values higher than 20% as well as PV systems exhibiting a nighttime production are considered as unrealistic and are therefore also removed from the original dataset.

3   Smart Meter Based Forecasting

A wide variety of methods are suggested in the scientific literature concerning time series based prediction. ANNs are nevertheless considered among the most successful machine learning algorithms for this purpose [6,7]. A feed-forward Multilayer Perceptron (MLP) available in the cloud computing software H_2O [8] is used in this study and deployed in local mode on the Azure VM. Concerning the network architecture, one hidden layer consisting of 200 neurons appears to be a good trade-off between accurate predictions and reasonable computational time. The rectifier $max(0, x)$ serves as an activation function, notably showing a higher performance than the Sigmoid function for individual SM profiles and low levels of aggregation. Furthermore, a random 50% of incoming weights are zeroed out to prevent overfitting and stochastic gradient descent with backpropagation is used to train the model with a prediction horizon of 24 h, i.e. 96 time steps, starting at midnight. It is assumed that all SM data until midnight is available for the model training and validation. Regarding the meteorological data, values recorded at the same time as the energy value to be predicted are used. This presupposes, though, a perfect weather forecast, which is certainly impossible in reality. The potential impact on the forecasting accuracy is discussed in more detail below, where the presented ANN is assessed on the three different types of datasets.

3.1   Small Consumers

Feature Selection. In this dataset, 27,284 profiles of residential loads, shops, small offices and a few electric storage heaters remain after the preprocessing tasks described in the previous section. Data from April 2014 to March 2015, i.e. one entire year, builds the training set while the month of April 2015 serves as the validation set. Furthermore, four types of data are gathered and used as input features for the neural network. The SM time series itself is the first source of information, from which 16 features are extracted as suggested, in part, by Valtonen et al. in [9]:

- Mean consumption of previous day,
- Last 3 values of previous day,
- Consumption on previous day at the same time, and 3 preceding time steps,
- Average of 3 previous days at the same time, and 3 preceding time steps,
- Average of 3 previous weeks on the same weekday at the same time, and 3 preceding time steps.

Note that multiple consecutive time steps are presented to the ANN simultaneously in order to make use of the temporal structure provided by the time series data. This allows the model not to rely only on a single value that can vary considerably from one time step to the next but to detect a consumption tendency at the considered time period. Instead of a standard feed-forward network, a variant of Time Delay Neural Network (TDNN) is employed, which is known to outperform the simple version [10].

Additionally, three different types of exogenous information are used to train the neural network, which can greatly increase the forecasting accuracy. Concerning weather data, only air temperature is considered in this paper. Since this feature appears to have a limited influence on the prediction performance, errors in the temperature forecast would not significantly impact the result. Another category consists of calendar features such as the hour of the day, the weekday and the month. The energy consumption depends finally to a large extent on social activities. For instance, at a household scale, the start-up of a single device like an electric oven or a washing machine is clearly visible in the load profile. Although it is inconceivable to accurately measure the human activity, one can still account for public holidays. To summarize, 21 features are fed to the ANN, from which a large share directly comes from historical energy values.

Spatial Aggregation. Besides training a model on single time series, different aggregation sizes are investigated, i.e. the aggregation of 10, 100, 1,000, 10,000 and all load profiles. Two options can be implemented which are presented in Fig. 1 with an example of 10 SMs:

(a) The forecast is carried out for each single profile before building groups of 10 randomly chosen predicted time series and adding them up,
(b) Original time series are first added up according to the previous group formation and the forecasting algorithm is only applied to aggregate load profiles.

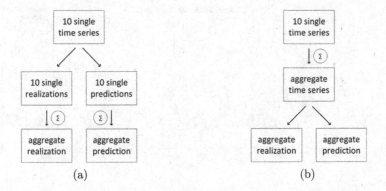

Fig. 1. Two versions to perform prediction with an aggregation of 10 load profiles.

Load Profiles. As previously mentioned, the model is trained during a whole year and evaluated in April 2015, which requires 2 to 4 seconds per SM. Figure 2 illustrates the case of an individual household during the fourth week of April. The predicted profile in red roughly follows the main tendency but fails to fit in with all spikes exhibited by the household, which is in fact hard to predict due to the intermittent nature of home appliances. Since a majority of SMs in this dataset record residential customers with a low and mostly unforesee-able energy consumption, the prediction algorithm inevitably performs poorly on average as described quantitatively in the following subsection. However, one can expect that the performance improves when time series of multiple house-holds are added up because the aggregation exhibits smoother and more seasonal patterns. Figure 3 displays both alternative options (shown in Fig. 1) to com-bine forecasting and aggregation of 100 load profiles which have been randomly selected from the SM dataset. According to the load shape, this sample consists of a large portion of late evening consumers, typical of households. Here, the ANN overestimates the real consumption for some business days during daytime and notably neglects the brief high spike on Saturday but generally succeeds in estimating the mean load profile hour after hour. Although shops and small offices are a minority among the subsets from 1,000 SMs on, their comparatively larger energy requirements during business hours usually increase the aggregate electricity demand up to the evening peak load level. In this case, the prediction algorithm does fairly well during working days but underestimates the amount of energy consumed at the weekend. The ANN probably places too much faith in features relating to previous days instead of giving weight to weekday based features. Finally, similar smooth profiles and successful outcomes appear for aggregations of 10,000 and of all time series as depicted in Fig. 4. Note that the variant where time series are added up in a second phase typically tends to yield a prediction that shrinks the real load profile due to the smoothing effect of the forecasting algorithm, which is notably visible at high aggregation levels.

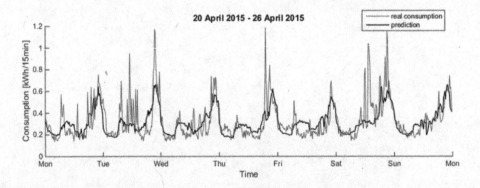

Fig. 2. Forecasting outcome of a single residential load.

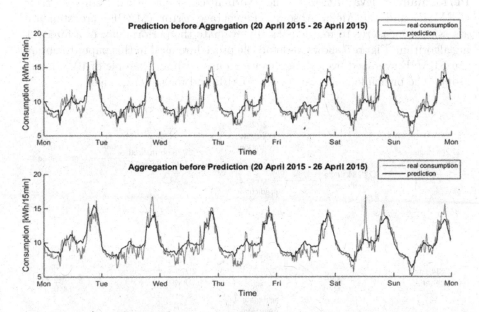

Fig. 3. Forecasting outcome of a random aggregation of 100 load profiles.

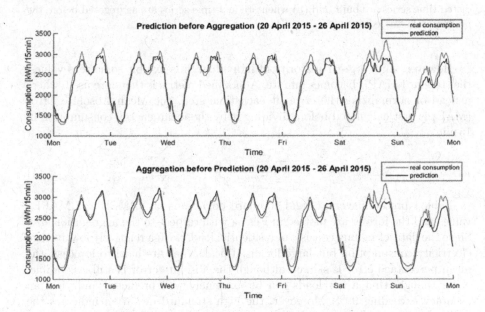

Fig. 4. Forecasting outcome of the sum of all available load profiles.

Performance Evaluation. The Normalized Root Mean Square Error (NRMSE) and the Mean Absolute Percentage Error (MAPE) are standard assessment measures to mathematically evaluate the performance of a forecasting algorithm. Figure 5 shows in detail the procedure used in this paper to obtain the NRMSE averaged over all aggregate groups with an example of 10 SMs per group. The principle is similar for the MAPE and for larger groups.

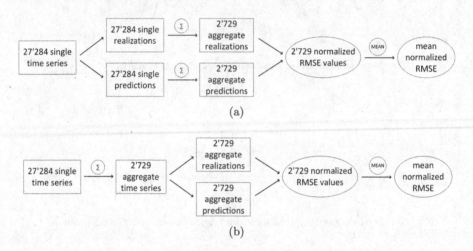

Fig. 5. Computation of the mean NRMSE (a) when groups of 10 individually predicted time series are built and (b) when original time series are aggregated before the prediction is carried out.

However, since a great majority of individual SMs exhibit some zero values, the standard MAPE becomes infinite. A modified metric is therefore used, where instead of normalizing the error at each time step, the Mean Absolute Error (MAE) is first calculated before dividing it by the mean energy consumption in 15 min:

$$\text{MAPE}^* \quad = \quad \frac{\frac{1}{m_{\text{test}}} \sum_{i=1}^{m_{\text{test}}} |e_i|}{\frac{1}{m_{\text{test}}} \sum_{i=1}^{m_{\text{test}}} y^{(i)}} \quad = \quad \frac{\sum_{i=1}^{m_{\text{test}}} |e_i|}{\sum_{i=1}^{m_{\text{test}}} y^{(i)}} \tag{1}$$

Table 1 presents averages and standard deviations of NRMSE and MAPE* values and the former are plotted in Fig. 6a with respect to the aggregation size. Since the dataset mainly consists of residential loads with a relatively low average electricity consumption but large localized peaks that are hard to forecast, the mean percentage error is skewed although the absolute error is still acceptable. Note, though, that a few loads can be extremely well predicted and yield an accuracy exceeding 99%. Moreover, the high standard deviation indicates the large variety of single consumers. Nevertheless, better predictions are made with larger groups, which can also be seen graphically in Figs. 2, 3 and 4. From an aggregation size of 1,000 SMs and larger, the ANN performance remains more or less constant. Since many studies related to short-term load forecasting by means of SM data are based on hourly measurements, it would not be judicious to compare them with results acquired in this paper. In addition, the efficiency of a prediction algorithm highly depends on the quality and the type of input data. Nevertheless, an analogous study using 15-minute energy values with aggregation sizes between 20 and 400 consumers, published by SAP Research, achieves a similar level of performance based on a Seasonal Naïve (SN) algorithm [11].

Table 1. Averages and standard deviations of NRMSE and MAPE* for different spatial aggregation levels for both aggregation-prediction variants.

Group size	mean (NRMSE)	std (NRMSE)	mean (MAPE*)	std (MAPE*)
Single SM	264.8	8263	166	6726
10 SMs (a)	47.6	25.6	33.6	20.8
100 SMs (a)	15.0	2.5	11.1	1.8
1,000 SMs (a)	7.9	0.5	5.9	0.3
10,000 SMs (a)	6.8	0.2	4.9	0.1
27,284 SMs (a)	6.7	-	4.9	-
10 SMs (b)	42.7	12.4	29.4	8.2
100 SMs (b)	15.4	2.3	11.4	1.7
1,000 SMs (b)	8.1	0.6	5.8	0.4
10,000 SMs (b)	6.6	0.06	4.4	0.03
27,284 SMs (b)	6.9	-	5	-

Examining the results of Table 1, a surprising fact is the almost identical forecasting accuracy of versions (a) and (b) for any aggregation size. However, the performance of the two versions is uneven when considering the difference between the measured and the forecasted demand averaged over all evaluation days and all consumers. While the real mean consumption is underestimated for any aggregation level as shown by negative values in Fig. 6b, the predicted amount of energy per day is closer to the reality in variant (b), where the energy difference also varies with the aggregation level, in contrast to the other version. Notably, by looking at the load profile examples of Fig. 4, variant (a) frequently underestimates the demand with the exception of a few night hours whereas variant (b) tends to offset shortfalls at night by surpluses during peak load time.

Instead of grouping randomly chosen SMs, an alternative option is to aggregate loads connected to the same Data Concentrator (DC), i.e. located in the same neighborhood. The City of Basel is actually equipped with more than 400 DCs that each collects energy information from a couple to several hundred SMs before forwarding the data to the central server of IWB. With the aim of increasing the forecasting accuracy, all consumers are first clustered into 5 groups by means of the K-Means algorithm provided by H_2O [8]. The Euclidean

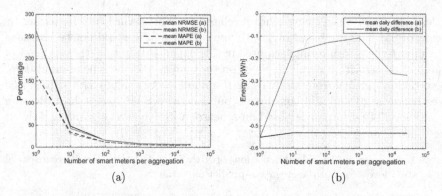

Fig. 6. Performance evaluation of ANN models with respect to the level of aggregation for both prediction-aggregation combinations: (a) average of NRMSE and MAPE*, (b) mean daily energy difference averaged over all consumers.

distance between clustering features such as the typical weekly profile as well as the mean consumption determines the cluster formation. The number of clusters K results from the Elbow method as illustrated by Fig. 7a, which suggests to choose K as the cluster number where the K-Means objective function exhibits an elbow. Consequently, load profiles related to the same DC and that belong to the same cluster are first aggregated together, then the forecast is carried out individually for these aggregate time series which are finally added up to build one profile per DC. In this case, data is trained from April 2014 to March 2015 and tested between April and August 2015. Figure 7b shows the performance at each DC and highlights the considerable dependency of the prediction performance on the energy requirements. Submitting an aggregation of similar time series to the neural network allows to achieve a median MAPE* of 13.06% or a median NRMSE of 18% per DC, and a performance of 3.36% (MAPE*), respectively 5.26% (NRMSE), when considering the sum of all available load profiles.

3.2   Commercial and Industrial Loads

Large consumers are characterized by a higher demand and more periodic patterns, which is reflected by a generally better forecasting accuracy. The dataset considered in this paper consists of 832 commercial and industrial load profiles that have successfully passed through the preparation process. The same features as in the case of small consumers are extracted from individual time series to train the ANN between May and August 2015 and evaluate the model on the data from September 2015. In this case, a median MAPE* of 12.6% and a median NRMSE of 17.6% per large consumer are obtained. Note that the forecast of the sum of all profiles achieves an excellent MAPE* performance of 1.96%.

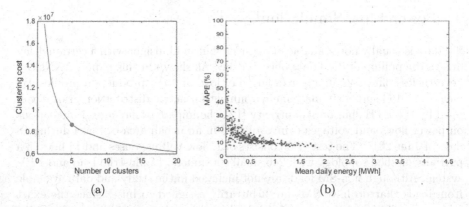

(a)   (b)

Fig. 7. (a) Elbow method leading to the choice of 5 clusters, (b) MAPE* associated to each DC with respect to the mean daily energy consumption.

3.3 Photovoltaic System Power Production

The same methodology can be used to forecast the PV production even if different input features are required. In particular, it appears that the accuracy gets better without historical SM data. Meteorological data such as global solar radiation, air temperature, atmospheric pressure and relative air humidity are on the other hand of great importance to train the ANN. Hence, weather forecast errors, especially concerning solar radiation, can considerably deteriorate the prediction performance such that presented outcomes must be considered with caution. Furthermore, the hour of the day and the month complete the set of exogenous features. The ANN is then trained from May to July 2015 and evaluated in August 2015. For the sake of consistency, negative predicted values are replaced with zero. Figure 8 illustrates the result during the first 7 days of August for a PV system exhibiting a 9.1% MAPE*. Most PV systems show very similar production patterns such that the performance mainly varies with the corresponding nominal power. In this dataset, a median MAPE* of 13.2% and a median NRMSE of 25.7% are obtained. The forecast based on the sum of all time series finally leads to a result similar to Fig. 8 and a MAPE* of 8.4%.

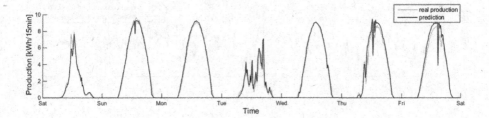

Fig. 8. Forecasting outcome of a PV system.

4   Power Flow Simulation

SM data is usually not available in real-time but in hindsight with a certain delay
due to the polling of data only once per day. As shown in this paper, ANNs can
nevertheless offer a good approximation of load and PV production profiles. By
means of DPG.sim, a simulation environment for active distribution grids developed by the ETH spin-off Adaptricity [12], the impact of day-ahead predictions
on power flows and voltages is investigated in an urban area of Basel during 18
days of June 2015. The test grid consists of 12 low-voltage buses and 14 lines in a
partially meshed topology with 246 small consumers, 1 industrial load and 1 PV
system. Although non-SM loads are not included in the study and data available
from loads that do have SMs are arbitrarily assigned to buses since the exact
address is unknown for privacy reasons, the case study can still be considered as
realistic. Furthermore, all loads are assigned with a fixed inductive power factor
of 0.97. The prediction is carried out first for each individual time series, which
gives a median MAPE of 16.33% for the active load per bus. As illustrated by
Fig. 9a for one of the lines, simulated active power flows based on measured and
on forecasted SM profiles are similar and the median MAPE of 17.88% is comparable to prediction accuracy of the active load per bus. In addition, Fig. 9b
reveals that bus voltages are barely modified (median MAPE = 0.032%). Note,
though, that the voltage is not really sensitive to active power injections. These
promising results still need to be validated with other grid topologies and different SM profiles but potentially show that forecasts based on SMs can provide
DSOs with additional valuable information, notably for real-time grid operation.

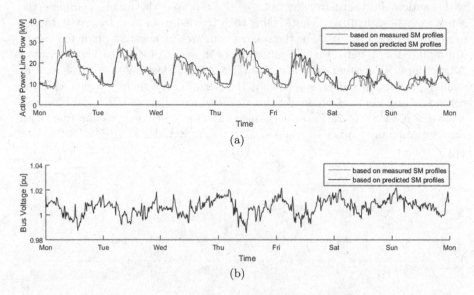

Fig. 9. Simulation of (a) active power line flow (MAPE = 13.67%) and (b) bus voltage
(MAPE = 0.029%) based on measured and predicted SM profiles.

5   Conclusion

Based on a large database, this paper proposes an exhaustive approach for forecasting various types of SM profiles, from an appropriate data preparation to the use of predicted time series in a power flow study. While an individual household is difficult to forecast, a considerably improved accuracy is achieved for commercial and industrial loads, PV systems and aggregate load profiles. Furthermore, training an ANN on spatially aggregated time series instead of adding up individual predicted profiles allows to reduce the computational cost without decreasing the forecasting accuracy. An even better efficiency can still be obtained by aggregating profiles of similar shapes. It would be nevertheless worthwhile to consider longer training and validation periods and investigate other prediction algorithms on this type of data, e.g. Support Vector Machine (SVM) or more sophisticated ANNs like Recurrent Neural Network (RNN) and Long Short-Term Memory (LSTM). In addition, the impact of weather forecast errors should be closely assessed. Eventually, based on satisfactory SM predictions, DSOs would be able to gain insight into the state of their low-voltage grid in real-time even though real-time measurements are not directly available.

Acknowledgements. This study is part of the project "Optimized Distribution Grid Operation by Utilization of SmartMetering Data" funded by the Swiss Federal Office of Energy and carried out at the ETH Zurich in collaboration with the ETH spin-off company Adaptricity [12] and the public utility of the City of Basel "Industrielle Werke Basel" (IWB).

References

1. Industrielle Werke Basel. http://www.iwb.ch/
2. MeteoSwiss. http://www.meteoswiss.admin.ch
3. Troyanskaya, O., Cantor, M., Sherlock, G., Brown, P., Hastie, T., Tibshirani, R., Botstein, D., Altman, R.B.: Missing value estimation methods for DNA microarrays. Bioinformatics **17**(6), 520–525 (2001). Oxford University Press
4. Alcalá-Fdez, J., Sánchez, L., García, S., Del Jesus, M., Ventura, S., Garrell, J., Otero, J., Romero, C., Bacardit, J., Rivas, V., Fernández, J., Herrera, F.: Knowledge Extraction Based on Evolutionary Learning. http://sci2s.ugr.es/keel/algorithms.php
5. Apache Spark. https://spark.apache.org
6. Koponen, P., Mutanen, A., Niska, H.: Assessment of some methods for short-term load forecasting. In: IEEE PES ISGT Europe (2014)
7. He, W.: Deep neural network based load forecast. Comput. Model. New Technol. **18**(3), 258–262 (2014)
8. OxData H2O. http://www.h2o.ai
9. Valtonen, P., Honkapuro, S., Partanen, J.: Improving short-term load forecast accuracy by utilizing smart metering. In: CIRED Workshop, Lyon, France (2010)
10. Busseti, E., Osband, I., Wong, S.: Deep learning for time series modeling. In: CS 229 Final Project Report. Stanford University (2012)
11. Ilic, D., Karnouskos, S., Da Silva, P.G.: Improving load forecast in prosumer clusters by varying energy storage size. In: IEEE Grenoble PowerTech (2013)
12. Adaptricity AG. https://www.adaptricity.com

Cybersecurity for Smart Cities: A Brief Review

Armin Alibasic, Reem Al Junaibi, Zeyar Aung$^{(\boxtimes)}$, Wei Lee Woon,
and Mohammad Atif Omar

Institute Center for Smart and Sustainable Systems (iSmart),
Masdar Institute of Science and Technology, Abu Dhabi, United Arab Emirates
{aalibasic,raljunaibi,zaung,wwoon,momar}@masdar.ac.ae

Abstract. By leveraging advancements in information and communications technology (ICT), Smart Cities offer many potential benefits like improved energy efficiency, management and personal security. However, this dependence on ICT also makes smart cities prone to cyber attacks. In this paper, we investigate the topic of cybersecurity for smart cities. We show how the specific characteristics of smart cities give rise to cybersecurity challenges, and review the different threats faced. Finally, we review some of the more important cybersecurity solutions for smart cities that have been proposed.

Keywords: Cybersecurity · Cyber threats · Smart city · Smart grid · Internet of Things

1 Introduction

Rapid advances in ICT have been exploited to streamline the design, operation and management of urban environments in a variety of ways. For example, it is now possible to monitor and manage energy consumption patterns in real time with smart meters, use this information to coordinate generation and distribution resources via the smart grid, continuously track traffic congestion and road hazards and communicate this automatically to vehicles and commuters. Progress in these areas have helped to cut costs, increase efficiency, bring about greater safety and convenience, and reduce pollution and greenhouse gas emissions.

A smart city can be defined as: *"A smart city uses digital technologies or information and communication technologies (ICT) to enhance quality and performance of urban services, to reduce costs and resource consumption, and to engage more effectively and actively with its citizens. Sectors that have been developing smart city technology include government services, transport and traffic management, energy, health care, water and waste."* [23].

According to well-known urban strategist Boyd Cohen, smart cities can be divided into six key components: (1) Smart Economy, (2) Smart Environment, (3) Smart Government, (4) Smart Living, (5) Smart Mobility, and (6) Smart People. Table 1 shows the six key components for smart cities [12].

© Springer International Publishing AG 2017
W.L. Woon et al. (Eds.): DARE 2016, LNAI 10097, pp. 22–30, 2017.
DOI: 10.1007/978-3-319-50947-1_3

Table 1. Key smart city components

Component	Indicators and benefits
Smart economy	Entrepreneurship & innovation, productivity, local and global interconnectedness
Smart government	Supply and demand side policies, ICT, e-government application, transparency, open data
Smart living	Culturally vibrant, happiness, health and safety
Smart mobility	Connected, ICT, support for clean and non-motorized options, mixed modalities
Smart people	Creative, inclusive, emphasis on educational excellence

While many of the "flagship" smart city developments have been designed from the ground up, the concept stands to make the most impact in situations where ICT technology is progressively integrated into the operations of existing urban areas. Cities can become "smart" by adopting modern technologies for transportation, traffic control, disaster response and security, resource management and other aspects of city management.

These enhancements are extremely valuable do carry a number of inherent risks. These tend to be rooted in the fact that a smart city often entails many new systems and devices being deployed in novel circumstances and often without adequate security testing. Many of these technologies are wireless and so depend on custom protocols and encryption platforms. Even seemingly minor bugs can cause very serious problems. For example in May 2012 the placer county courthouse system in California accidentally summoned 1,200 people to jury duty on the same morning, an incident which resulted in severe traffic jams throughout the city [6]. In this particular case, the event was the result of an unintentional computer glitch, but it would not be difficult to envision a situation where a hostile party could intentionally create a similar "glitch" to disrupt public life in a similar way.

Of even greater concern is the fact that many smart cities have yet to develop action plans which outline responses to possible cyber attacks which target the city's services, infrastructure and ICT systems. Because all systems are fundamentally interconnected, weaknesses in any one element can have wide-ranging consequences. For example, encryption problems can result in a compromised wireless network, which in turn can be exploited by hackers to attack a city's electricity or water supply. It is clear that cyber threats to smart cities need to be taken extremely seriously.

Possible solutions include:

1. Creation and use of security check lists for encryption, authentication, authorization, and software updates while implementing new systems
2. Implementation of failsafes and manual overrides on all city systems
3. Development of action plans and procedures for responding to cyber attacks

We will focus more on these solutions in the evaluation section of this paper. According to Gartner [5], by the end of 2020 there will be 25 billion connected devices. Other projections [4] indicate that 70% of the world's population is expected to live in urban environments by 2050. The rapid growth in the world-wide urban population, as well as the increasing interconnectedness of this demographic makes the cybersecurity of cities incredibly important, and the situation will only become even more pressing with the introduction of ever more intelligent and connected devices and infrastructure.

A wide variety of systems, ranging from home appliances to medical devices in hospitals to air defense systems, will be affected by a single cyber attack which targets the energy grid. The weapon of choice in this modern age is no longer a bomb, but rather malicious software (known as malware) designed to destroy, disrupt or take control of the complex systems which control the operation of smart grids. In addition, the immense complexity and scale of a smart city would mean that these issues need to addressed as early as possible, and that, in the case of many existing cities, it may already be too late to incorporate adequate cybersecurity measures.

The cyber threat landscape is extremely fluid. The last few years have seen an exponential growth in the number of potential threats. In a 2012 report, McAfee Labs stated that, *"For the year, new malware sample discoveries increased 50 percent with more than 120 million samples now in the McAfee Labs 'zoo' "* [17]. The specific nature of the threats themselves are also evolving and are increasing in sophistication. Advanced persistent threats (APT), where an unauthorized entity gains and retains access to a network, are a good example of this trend. In many cases, the attackers are no longer "script kiddies", but are highly skilled and organized professionals who are able to deploy a variety of sophisticated techniques to launch complex and coordinated attacks. Examples of well-known cyber threats include:

- Hackers
- Malware
- Zero days
- Botnets
- Denial of service (DOS)
- Distributed denial of service (DDOS)

While all these terms are by now quite widely known, the scope of the attacks have now broadened and include industrial control systems, as was demonstrated in 2010 with Stuxnet [24].

2   Smart City Cyber Challenges

There is an extensive body literature on the topic of smart cities. Some papers [13,18] provide useful guidelines for policy makers and city managers seeking to better define and drive their smart city strategy and planning actions towards the most appropriate domains of implementation. Other papers [19,25] have described the deployment and experimentation architecture of the Internet of Things (IoT) so they can provide a suitable platform for large scale experimentation and evaluation of IoT concepts under real-life conditions. Some authors [14] apply a Quality Function Deployment (QFD) method to establish interconnections between services and devices, and between devices and technologies for smart cities.

However none of above mentioned papers emphasized cybersecurity. Seto et al. [20] discussed the privacy risks associated with advances in the standardization of the smart grid, whose technology is at the core of the smart city. They verified the effectiveness of privacy impact assessment, with reference to privacy risks in the smart city, where all kinds of user data are stored in electronic devices, thus making everything intelligent. Yibin et al. [15] presented a mobile-cloud-based smart city framework, which is an active approach to avoid data over-collection. By putting all of the users' data into the cloud, the security of users' data can be greatly improved.

Matuszak et al. [16] and Wang et al. [22] carried out a series of studies on reducing the risks of cyber intrusions and detecting various types of attacks on the smart grid, which can be regarded as a subset of the smart city, and developed algorithms and visualization techniques for cyber trust in a smart grid system. Cyber trust was evaluated in terms of a mathematical model consisting of availability, detection and false alarm trust values, as well as a model of predictability.

In addition to the above theoretical studies, researchers have also performed experiments on real-world scenarios. Research conducted by Hioureas and Kinsey [11] proved how surveillance technology systems could also be used in a harmful way. They performed man-in-the-middle attack by using methods such as Address Resolution Protocol (ARP) poisoning. That enabled them to alter any data sent to and from the router. Figure 1 shows an attacker who tells the user he is the router, and tells the router he is the user, thus intercepting traffic to and from the web server is easy.

Fig. 1. An attacker masquerades as the router to gain access to sensitive communication

Cerrudo [7] provided an overview of current cybersecurity problems affecting cities as well as real threats and possible cyber attacks that could have a huge impact on cities. Some of possible weak points mentioned were:

- Traffic Control Systems
- Smart Street Lighting
- City Management Systems
- Sensors
- Public Data
- Mobile Applications
- Cloud and Software as a Service (SaaS) Solutions
- Smart Grid
- Public Transportation
- Cameras
- Social Media
- Location-based Services

One example that was mentioned pertained to an attack on sensors, which form the backbone of the smart city: "*Attackers could even fake an earthquake, tunnel, or bridge breakage, flood, gun shooting, and so on, raising alarms and causing general panic. An attacker could launch a nuisance attack by faking data from smell or rubbish level sensors in empty garbage containers, to make garbage collectors waste time and resources. Keep in mind that many systems and services from cities rely on sensors, including smart waste and water management, smart parking, traffic control, and public transport. Hacking wireless sensors is an easy way to remotely launch cyber attacks over a city's critical infrastructure.*" [7].

According to a poll conducted by researchers at the Morning Consult firm [2], nearly 36 percent of voters consider acts of terrorism as the main security threat to the USA, followed by cyber-attacks at 32 percent, while "war with a large military power" was perceived as the third greatest threat with 12.1% of the vote (see Fig. 2). With all this in mind and from analyzing the papers cited above, we can see the importance of devoting additional resources and attention to securing smart cities from cyber attacks.

3   Proposed Solutions

One of many possible smart city cybersecurity solutions was proposed by Cerrudo et al. [8]. In their report, the authors provided guidelines for the organizations responsible for selecting and testing the technologies which would be deployed in a smart city. The focus was on appropriate testing and assessment strategies to be followed when selecting these technologies as well as the respective vendors.

Firstly, the importance of a structured and well thought out technology selection process cannot be understated. In particular, it must be stressed that issues of cybersecurity should be taken into account right from this early stage. In the context of a smart city, all systems are inter-dependent and weak services can cause large-scale damage and even affect national stability and security. A smart

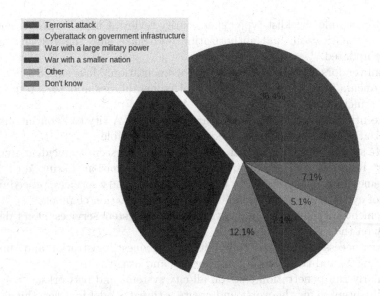

Fig. 2. Perceived US security threats [3]

city requires new levels of confidentiality, integrity, availability and defense. All wired and wireless communications (data in transit) should be properly protected with strong encryption. The solution should support strong authentications mechanism (one-time passwords, certificate or biometric-based authentication etc.). All functionality should require and enforce proper permissions (authorization) before performing any action. Updates of software, firmware, etc. should be automatic and secure. Logs must also be saved securely against tampering. Devices should have a mechanism to prevent tampering by unauthorized sources. In the case of a system malfunctioning or crash, the system should remain secure and security protections remain enforced. Solutions should come with a secure configuration by default. All of these their recommendations are for Technology Selection.

The second is the recommendations for technology implementation, operation and maintenance. For implementation technology should pass selection phase security test; technology should be securely delivered; enable strong encryption; secure system administration; set strong passwords; remove unnecessary user accounts; disable unused functionality and services; enable auditing of security events, etc. For operation and maintenance technology should pass monitoring, patching, regular assessment and auditing, protection of logging environment, access control, cyber-threat intelligence, compromise reaction and recovery.

The third is the recommendations for technology disposal. We should avoid repurposing technology, all data should be erased securely and if that is done by vendors then they should do the same.

Cerrudo et al. [8] also proposed a checklist of security related steps that smart city operators and administrators should consider implementing.

- Create a simple checklist-type cybersecurity review. Check for proper encryption, authentication, and authorization and make sure the systems can be easily updated.
- Ask all vendors to provide all security documentation. Make sure Service Level Agreements include on-time patching of vulnerabilities and 24/7 response in case of incidents.
- Fix security issues as soon as they are discovered. A city can continuously be under attack if issues are not fixed as soon as possible.
- Create specific city CERTs that can deal with cybersecurity incidents, vulnerability reporting and patching, coordination, information sharing, etc.
- Implement and make known to city workers secondary services/procedures in case of cyber attacks, and define formal communication channels.
- Implement fail safe and manual overrides on all system services. Don't depend solely on the smart technology.
- Restrict access in some way to public data. Request registration and approval for using it, and track and monitor access and usage.
- Regularly run penetration tests on all city systems and networks.
- Finally, prepare for the worst and create a threat model for everything.

Gurgen et al. [10] suggested smart city objectives should encourage self-awareness and provided a set of guidelines and recommendations to achieve this. One of the most frequently adopted models for realizing an autonomic system is the MAPE-K model (see Fig. 3), which consists of a control loop with four phases (Monitor, Analyze, Plan, and Execute) built on an underlying knowledge base,

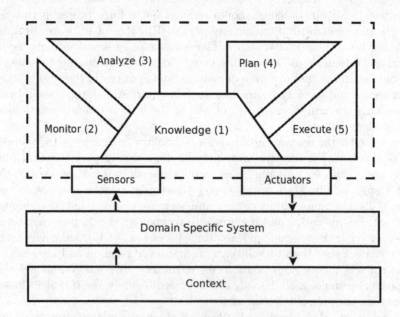

Fig. 3. The MAPE-K model [10]

and which interacts with the surrounding physical environment using sensors and actuators.

Dong [9] employed complex networks theory and data mining to identify vulnerabilities in the physical power system of a smart grid, which is a critical component in a smart city. The proposed cyber system models was designed to be used alongside existing power system models to analyze the complex interactions between the cyber and physical parts of a smart grid. The author also proposed advanced modeling tools to model cyber attacks and to evaluate how they could affect smart grid security.

For a broader review of information security issues, which encompasses all ICT-based systems, including the smart city, readers are referred to [21], which explains the fundamentals of information security in a very accessible manner. Topics addressed include CIA (Confidentiality, Integrity, and Availability), cryptography, cryptanalysis, access control, security protocols, and various aspects of software security.

4   Conclusion

In this paper, we discussed the concept of smart cities and their cybersecurity challenges and possible solutions. The area of smart city cybersecurity is still in its infancy, and many more policy, architectural, design, and technical solutions are anticipated in this very important domain. We would like to conclude this paper with the words of the renowned security expert Eugene Kaspersky [1]: "*Smart technologies and interconnectivity should be improving lives around the world. But with all the opportunities they create, it is a challenge to stop people with malicious intent exploiting them. we are confident that it is possible to meet the challenge, but it requires a lot of hard work from governments, software and equipment developers, and IT security companies. We are just starting out on this path, but follow it we must — to ultimately build a safe and secure digital world for all.*"

References

1. http://gulfnews.com/business/sectors/technology/cybersecurity-challenges-in-sma rt-cities-development-1.1613223
2. http://resources.infosecinstitute.com/cyber-attacks-on-power-grid-the-specter-of-total-paralysis/
3. http://securityaffairs.co/wordpress/wp-content/uploads/2015/07/power-grid-atta ck-scenario-2.jpg
4. https://www.qualcomm.com/products/smart-cities
5. http://www.gartner.com/newsroom/id/2905717
6. http://www.npr.org/2012/05/03/151919620/computer-glitch-summons-too-many-jurors
7. Cerrudo, C.: An emerging US (and world) threat: cities wide open to cyber attacks. Technical report, Securing Smart Cities (2015)

8. Cerrudo, C., Hasbini, A., Russell, B.: Cyber security guidelines for smart city technology adoption. Technical report, Securing Smart Cities, Cloud Security Alliance (2015)
9. Dong, Z.: Smart grid cyber security. In: Proceedings of the 2014 13th International Conference on Control Automation Robotics and Vision, pp. 1–2 (2014)
10. Gurgen, L., Gunalp, O., Benazzouz, Y., Gallissot, M.: Self-aware cyber-physical systems and applications in smart buildings and cities. In: Proceedings of the Conference on Design, Automation and Test in Europe, pp. 1149–1154 (2013)
11. Hioureas, V., Kinsey, T.: Does CCTV put the public at risk of cyberattack? How insecure surveillance technology is working against you. Technical report, Securing Smart Cities (2015)
12. Institute for Sustainable Communities, et al.: Getting smart about Smart Cities: USDN resource guide. Technical report, Institute for Sustainable Communities, Nutter Consulting, and Urban Sustainability Directors Network (2015)
13. Lazaroiu, G.C., Roscia, M.: Definition methodology for the Smart Cities model. Energy **47**, 326–332 (2012)
14. Lee, J.H., Phaal, R., Lee, S.H.: An integrated service-device-technology roadmap for Smart City development. Technol. Forecast. Soc. Chang. **80**, 286–306 (2013)
15. Li, Y., Dai, W., Ming, Z., Qiu, M.: Privacy protection for preventing data over-collection in Smart City. IEEE Trans. Comput. **65**, 1339–1350 (2015)
16. Matuszak, W.J., DiPippo, L., Sun, Y.L.: CyberSAVe: situational awareness visualization for cyber security of smart grid systems. In: Proceedings of the 10th Workshop on Visualization for Cyber Security, pp. 25–32 (2013)
17. Labs, M.A.: McAfee threats report: fourth quarter 2012 executive summary. Technical report, McAfee, Intel Security Group (2013)
18. Neirotti, P., De Marco, A., Cagliano, A.C., Mangano, G., Scorrano, F.: Current trends in Smart City initiatives: some stylised facts. Cities **38**, 25–36 (2014)
19. Sanchez, L., Munoz, L., Galache, J.A., et al.: SmartSantander: IoT experimentation over a Smart City testbed. Comput. Netw. **61**, 217–238 (2014)
20. Seto, Y.: Application of privacy impact assessment in the Smart City. Electron. Commun. Jpn. **98**, 52–61 (2015)
21. Stamp, M.: Information Security: Principles and Practice, 2nd edn. Wiley, Hoboken (2011)
22. Wang, P., Ali, A., Kelly, W.: Data security and threat modeling for Smart City infrastructure. In: Proceedings of the 2015 International Conference on in Cyber Security of Smart Cities, Industrial Control System and Communications, pp. 1–6 (2015)
23. Wikipedia.org: Smart City (2016). https://en.wikipedia.org/wiki/Smart_city
24. Wikipedia.org: Stuxnet (2016). http://en.wikipedia.org/wiki/Stuxnet
25. Yin, C., Xiong, Z., Chen, H., Wang, J., Cooper, D., David, B.: A literature survey on Smart Cities. Sci. China Inf. Sci. **58**, 1–18 (2015)

Machine Learning Prediction of Photovoltaic Energy from Satellite Sources

Alejandro Catalina, Alberto Torres-Barrán, and José R. Dorronsoro[✉]

Dpto. Ing. Informática, Universidad Autónoma de Madrid, Madrid, Spain
jose.dorronsoro@uam.es

Abstract. Satellite–measured irradiances can be an interesting source of information for the nowcasting of solar energy productions. Here we will consider the Machine Learning based prediction at hour H of the aggregated photovoltaic (PV) energy of Peninsular Spain using the irradiances measured by Meteosat's visible and infrared channels at hours $H, H-1, H-2$ and $H-3$. We will work with Lasso and Support Vector Regression models and show that both give best results when using $H-1$ irradiances to predict H PV energy, with SVR being slightly ahead. We will also suggest possible ways to improve our current results.

Keywords: Photovoltaic energy · EUMETSAT · SEVIRI channels · Lasso · Support Vector Regression · Nowcasting

1 Introduction

Global concerns on climate change, the extremely fast development of nations such as China and India and the obvious interest of clean and affordable energy are pushing forward the worldwide use of renewable energies, particularly solar. Thus, it is crucial to constantly improve PV energy production forecasting, the subject of a large research effort (see [1,6] for comprehensive reviews on recent work) and also be the focus of this work. PV energy prediction has two distinct horizons. For day ahead or longer prediction horizons, Numerical Weather Predictions (NWP), such as those of the European Center for Medium Weather Forecasts (ECMWF) are the key input. However, NWP are refreshed at best each 6 h with ECMWF runs starting at hours 00, 06, 12 and 18 UTC and which are widely available about six hours later. This means that, in Spain, the 00 run will be useful to predict PV energy for approximately the period from 05 UTC to 12 UTC and the 06 run for the period 12 UTC to 19 UTC; the forecasting improvement of the other runs would affect basically to night hours. Concrete hours in other countries will change but it is nevertheless clear that other inputs are needed for better short term or intraday forecasts of PV energy.

Section 7.2 in [1] gives a complete overview of several approaches for the intraday forecasting problem. A first approach is to work in an endogeneous, i.e. a pure time series (TS) scenario using for nowcasting purposes the latest PV energy

W.L. Woon et al. (Eds.): DARE 2016, LNAI 10097, pp. 31–42, 2017.
DOI: 10.1007/978-3-319-50947-1_4

readings; a relevant example here is [12] for a single PV plant in California, show-ing that the best results are obtained by a mixed Genetic Algorithm/Artifical Neural Network (GA/ANN) model. The pure time series information can be enlarged with other past information such as irradiances or temperatures [13] or combining it with NWP forecasts for the hours ahead, where a large number of contributions have been made. This approach was followed in [2] for the PV production of Peninsular Spain, where past readings are combined with the cor-responding NWP–based day ahead predictions emitted the day before to try to correct these day ahead predictions for the next hours. It was shown that this can improve on NWP–based day ahead forecasts for about 2–3 h, depending on the hour of the day. Images from sky cameras [8], Ch. 9, have also been consid-ered; for instance in [10] they are used for short term (up to 15') prediction of Direct Normal Irradiance (DNI). While very interesting locally, it is not clear how to exploit them over large geographic areas. This leaves us with satellite–based measurements [8], Ch. 3, that can offer at least in some of their channels relevant information with adequate update frequencies.

Satellite information has been used for nowcasting radiation values and PV energy having at its basis the HELIOSAT method to estimate solar irradiance from satellite images [4]. More precisely, images are first used to compute a dimensionless cloud index value which describes for a given hour H the influence of cloudiness on atmospheric transmittance; then this cloud index is used to estimate the ratio between actual global irradiance and the output of a clear-sky model, that can be computed for any given hour. Next, a Motion Vector Field approach is used to forecast cloud images for the following hours $H+1, H+2, \ldots$, which, in turn, allows to forecast cloud indices, irradiance ratios and, finally, irradiance values for these hours. While targeting initially irradiance forecasts, this approach has been also used to derive short term PV energy forecasts for PV plants and regional aggregations (see [7, 9, 15]).

In this work we are also going to use satellite images to derive short time PV forecasts but we will follow a different approach, purely Machine Learning (ML) based. We will have as the target the aggregated PV production of Peninsular Spain. Spain's PV landscape is quite fragmented, with 4,786.6 MW of installed power at the end of 2014, fairly concentrated in its southern half but with very few plants with peak power above 20 MW. This makes an individual plant–based approach to aggregated PV prediction hard to come by and therefore we will try to predict aggregated PV energy directly. Initially we will consider satellite images for METEOSAT's 11 spectral bands, i.e., from visible to long-wavelength infrared, with wavelength ranges 0.6 μm to 13.4 μm. METEOSAT also has a High Resolution Visible (HRV) channel; however we will not consider it here because of the homogeneity with other channels. In addition, its higher spatial resolution may not be so important in large area PV forecasts and its relevant information should also be captured by other visible channels considered.

Since current silicon PV cells cannot transform infrared rays, it is clear that infrared channels, particularly those farther away, won't be as relevant as the visible ones. Moreover, EUMETSAT measures reflected radiance which can only act as a proxy to the incoming irradiance that actually produces PV energy. Nevertheless, we will use channel information to predict PV energy at hour H from channel readings at hours H, $H-1$, $H-2$ and $H-3$. We will consider hourly PV and satellite data for the years 2013, 2014, and 2015, with 2013 used for model training, 2014 for validation when selecting model hyper-parameters and 2015 as the test set. We will use two different approaches, working first with linear Lasso sparse models and then with Gaussian kernel Support Vector Regression (SVR). SVR is of course more powerful and will indeed give better results but Lasso not only gives a base benchmark model but may allow, in principle, an easier model interpretation as its non-zero coefficients are associated to concrete geographic points. We can summarize our paper contributions as follows:

1. We analyze the capabilities and channel relevance of satellite measured irradiances to predict aggregated PV energy production over Peninsular Spain.
2. We discuss the application of two standard, widely used ML methods, Lasso and SVR, to predict PV energy up to 3 h ahead from the most relevant irradiance channel values.
3. We discuss model results and point out to ways to improve on them.
4. We give a first analysis of model interpretability, spatially for Lasso and temporally for SVR.

We emphasize that while we believe them interesting and useful, our results are to be seen as a first step towards a better use of satellite–measured irradiances in PV energy prediction. The paper is organized as follows. In Sect. 2 we will briefly review the EUMETSAT system, the satellite measures and other information it provides and the concrete channel information which can be obtained essentially in real time from EUMETSAT. These channels information is analyzed in Sect. 3 in the context of predicting the PV energy of Peninsular Spain. In principle it is to be expected that the shorter wavelength channels are the most related to PV energy and our analysis will confirm this. In turn, of these, the IR016, VIS008 and IR039 channels have the strongest correlation with PV energy and in Sect. 4 we will first study the quality of Lasso and SVR models to predict PV energy at a given hour H from the radiance measures at hours $H, H-1, H-2$ and $H-3$. As we shall see, best results are achieved when using irradiances at hour $H-1$, with SVR slightly outperforming Lasso. Finally, we will discuss further ways to improve on these results in Sect. 5.

2   Meteosat Satellite Data

The European Organisation for the Exploitation of Meteorological Satellites (EUMETSAT) operates a series of geostationary Meteosat satellites that cover, among others, Europe, Africa and the Atlantic Ocean. Their goal is to provide information on the radiance emitted and reflected by the Earth's surface and atmosphere to be used in meteorological forecasting, as well as in climate monitoring and research, often after further processing is done by the centers on EUMETSAT's Satellite Application Facility (SAF) network.

The Meteosat Second Generation (MSG) satellites are equipped with the Spinning Enhanced Visible and Infrared Imager (SEVIRI) technology and provide near real–time radiance values every 15 min over eleven spectral bands, going from the visible (with a wavelength 0.6 μm at Channel 1) to the long infrared (with a wavelength 13.4 μm at Channel 11); the spatial resolution is about 3 × 3 km for most of the covered regions. Some of these channels measure concrete properties, such as absorption of water-vapor (Channel 5), ozone (Channel 8) or CO2 (Channel 11). There is also a high-resolution channel with visible radiance on the 0.6 μm–0.9 μm wavelength range with a 1 × 1 km resolution over Europe and parts of Africa. Another product of interest is the Cloud Mask, a floating point number within the range [0, 5], where 0 means clear and 5 totally covered skies.

The basic pixel counts that SEVIRI collects over each channel are further processed by several calibration procedures. First, readings are converted into radiances assuming a linear relation, that is, by adding an offset to the pixel count and multiplying the resulting number by an appropriate calibration factor (as described in EUMETSAT's inter-calibration documentation). For the visible channels a reflectance percentage is computed as the fraction of their radiance to the maximum solar irradiance. The infrared radiances are converted into equivalent brightness temperatures through an empirical formula [7]. This means that the 11 channel radiances give rise to another 11 variable set, with 3 reflectances and 8 brightness temperatures. Thus, excluding the high resolution channel and adding the cloud mask we would have in principle 23 values at each grid point. Obviously, their use and interest will depend on the concrete objectives pursued.

This information has been usually applied to modeling solar irradiance at concrete locations, often after a detailed analysis of the physical processes involved. However, our goal, wide area PV energy forecasting, is quite different and while surface irradiance is certainly a key information, the very large number of installations involved, their different rated powers and their technological variations suggest to follow other paths. Here we will pursue an "agnostic", Machine Learning (ML) based approach, considering in principle all channel information as potentially relevant and focusing on some channels and discarding others not by physical reasons but by their information content with respect to the PV energy target. As we shall see, this results in discarding some a priori obvious channels but also in retaining others that at first glance would seem less relevant. We discuss next these issues together with the data we will use.

3   Satellite Data Analysis

We will use Meteosat channel readings at UTC hours 0 to 23 for the years 2013, 2014 and 2015 that we have downloaded from the EUMETSAT Data Centre. Given the large area we are concerned with, we have downsampled Meteosat's initial resolution to that of a 0.125° grid (the same resolution that has been standard until recently by the NWP forecasts of the ECMWF). We will work only with the grid points of Peninsular Spain; this results in a total of 3,391 points. We

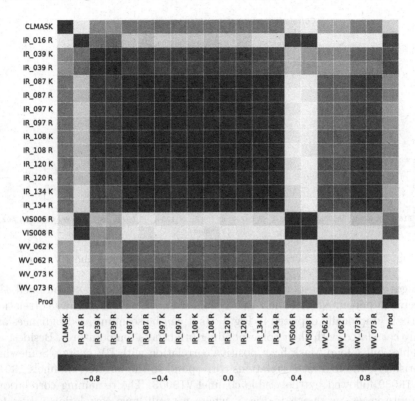

Fig. 1. Correlations between cloud index, satellital radiances and PV energy.

will consider first the information for all channels with hourly resolution, from January 1, 2013 to December 31, 2015. Although this should result in 26,304 h, the actual number is 25,161 h because of missing data. Of course, only daylight hours are relevant and we have simply selected at the 15-th day of each month the UTC sunrise and sunset hours of Girona (at the easternmost of Spain) and Santiago de Compostela (at the westernmost) respectively and rounded them to the closest hour (daily sunrise and sunset times usually vary by about 15' from those at the month's 15–th day). After dropping the non selected hours that we take as dark, we are finally left with 5,475 h per year.

Recall that we have in principle 23 satellite measures for each grid point that, if taken on their entirety, would result in input patterns with an extremely large dimension of $3,391 \times 23 = 77,993$. However, since our aim is PV energy nowcasting by Machine Learning methods, it is likely that the first visible or near visible channels would be of the greatest interest. Moreover, since their reflectance percentages are given essentially as dilations, they may not add useful information to the models. In any case, a basic measure of variable relevance is its correlation with the target variable, here PV energy.

To get a manageable first measure, we have computed the average of each variable at every hour over the grid points and then the correlation of these

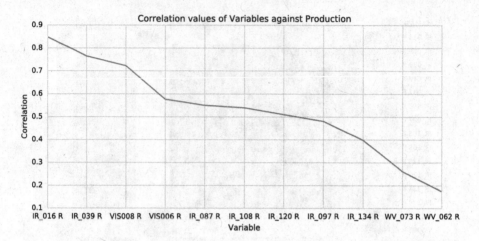

Fig. 2. Correlations between cloud index, satellital radiances and PV energy.

averages with the corresponding PV energy reading excluding, as mentioned, the reflectance percentages of the three visible channels. The resulting correlation matrix is graphically depicted in Fig. 1. As it can be seen, infrared radiances are highly correlated with their corresponding brightness temperatures. Besides, all variables but Cloud Mask have positive correlation with PV energy. Somewhat surprisingly, the highest correlations correspond to the infrared channels IR016 and IR039 followed by the visible channel VIS008. The remaining correlations are quite lower, as shown in Fig. 2, where we only give correlation values for radiances and sort channels by decreasing correlation.

Because of this, in what follows we will work only with the radiances of the IR016, IR039 and VIS008 channels plus the brightness temperature of channel IR039. Note that since channel data consists of radiances while PV data are energies, a possibly better analysis could be done considering for each hour the average radiances of the previous four 15' satellite readings. In any case, we recall that this is a first analysis to be extended in further work.

4   Modelling PV Energy from Channel Information

In this section we will use data from the selected channels to model PV energy at hour H with the radiance values over Peninsular Spain at $H, H - 1, H - 2$ and $H - 3$. To do so we will consider two well established ML models, Lasso and Support Vector Regression (SVR) which have received some recent attention in the literature [3,11,13]. We briefly describe them next.

4.1   The Lasso and Support Vector Regression Models

Given an N pattern sample $\{(x^1, y^1), \ldots, (x^N, y^N)\}$ with d–dimensional inputs x^p and 1–dimensional targets y^p, the Lasso solution [5] w^*, b^* minimizes the L_1 regularized loss

Table 1. Hyper–parameters of the Lasso and SVR m1N models.

Model	Parameter	mOM	mON	mOE	m1M	m1N	m1E	m2M	m2N	m2E	m3M	m3N	m3E
Lasso	λ	0.033	0.017	0.027	0.022	0.023	0.055	0.106	0.065	0.040	0.157	0.073	0.036
SVR	$C\left(\times 10^3\right)$	0.170	0.155	0.178	0.321	0.474	0.096	0.184	0.109	0.054	0.102	0.090	0.049
	ϵ	0.034	0.112	0.019	0.123	0.026	0.042	0.132	0.075	0.032	0.053	0.168	4.117
	$\gamma\left(\times 10^{-5}\right)$	3.064	3.085	3.061	3.060	3.053	3.350	3.054	3.057	4.683	3.770	3.069	6.108

$$\ell_L(w,b) = \frac{1}{2N}\sum_p (w \cdot x^p + b - y^p)^2 + \lambda\|w\|_1.$$

The L_1 regularization introduces sparsity in the model with two consequences. First, it avoids possible singularities in the sample covariance matrix; notice that sample size, 5,475, is smaller than input dimension, $13,564 = 3,391 \times 4$. Second, L_1 regularization drives many model coefficients to zero, and the positions of the non-zero coefficients may yield first interpretation of the model.

An obvious drawback of Lasso is its possibly poor modeling results because of its linear nature. This motivates our choice of the non linear and possibly more powerful Gaussian kernel SVR [14], one of the workhorses in non linear regression, as our second model. Assuming first for simplicity a linear SVR model, the SVR cost function is

$$\ell_S(w,b) = \sum_p [y^p - w \cdot x^p - b]_\epsilon + \frac{1}{C}\|w\|_2^2; \tag{1}$$

here we use now the ϵ-insensitive loss $\ell(y,\hat{y}) = [y - \hat{y}]_\epsilon = \max\{|y - \hat{y}| - \epsilon, 0\}$ and L_2 regularization. We thus allow an ϵ-wide, penalty-free "error tube" around the model. To find the optimal w^*, b^*, (1) is rewritten as a constrained minimization problem which is then transformed using Lagrangian theory into the dual problem, the one actually solved; see [14] for more details. To improve on a linear model, the kernel trick [14] is used to take advantage of the fact that only dot products are involved when solving the dual problem. This allows to replace the initial products $x \cdot x'$ with a positive definite kernel $k(x, x')$ that can be written as $k(x, x') = \phi(x) \cdot \phi(x')$, where the x are mapped through $\phi(x)$ into a larger, possibly infinite dimensional Hilbert space H. We obtain thus a non linear model $f(x) = W \cdot \phi(x) + b$ and, in turn, the optimal $W^* \in H$ can be written as $W^* = \sum \alpha_p^* \phi(x^p)$, where the x^p for which $|\alpha_p^*| > 0$ are the Support Vectors (SVs). We thus have

$$f(x) = b^* + W^* \cdot \phi(x) = b^* + \sum \alpha_p^* \phi(x^p) \cdot \phi(x) = b^* + \sum \alpha_p^* k(x^p, x).$$

Using a Gaussian kernel $e^{-\gamma\|x-x'\|^2}$ gives a model $f(x) = b^* + \sum_p \alpha_p^* e^{-\gamma\|x-x^p\|^2}$. Note that the SVR model also lends itself to an interpretation from a temporal point of view, as the SVs correspond to the day–hour pairs whose radiances define the centers of the different Gaussians the model is made of.

Both models require careful hyper–parameter selection, λ for Lasso and C, ϵ and γ for SVR, which we will do using the year 2013 as a training set and 2014

Table 2. Lasso and SVR monthly average test errors for year 2015.

Month	H		H-1		H-2		H-3	
	Lasso	SVR	Lasso	SVR	Lasso	SVR	Lasso	SVR
January	3.39	2.83	2.42	2.31	4.37	3.53	6.42	5.17
February	4.17	3.65	3.37	3.15	4.48	4.14	6.83	6.77
March	3.98	3.46	3.38	3.01	4.91	4.79	6.87	6.30
April	4.14	3.48	3.80	3.00	4.80	3.29	5.37	4.38
May	3.16	2.33	2.80	2.36	3.84	2.87	4.11	3.43
June	2.99	2.71	2.57	2.47	3.26	2.45	3.64	2.99
July	2.19	2.48	2.41	1.97	3.33	2.20	4.03	2.62
August	3.09	2.80	2.90	2.67	3.09	3.00	3.66	3.28
September	3.03	2.82	2.87	2.58	3.94	3.14	4.49	3.86
October	3.18	3.63	3.12	3.67	4.10	4.28	5.36	5.71
November	2.98	3.32	2.20	2.94	3.86	3.62	6.12	5.90
December	3.39	4.07	2.75	3.47	4.29	4.39	6.00	5.81

as the validation set. Notice that the data have a strong temporal structure that might be partially lost if standard k–fold CV were used. On the other hand, error estimates over an entire year should be robust and significant enough.

4.2 Lasso and SVR Results

Recall that we will predict PV energy for an hour H using four variables at each grid point from hours $H, H - 1, H - 2$ and $H - 3$, which we will indicate with the indices 0, 1, 2 and 3. In other words, we consider four problems according to the distance between the hour for which PV energy is sought and the hour from which readings are used. While in principle we could use a single model for all day hours, notice that this could work fine for the H problem, but it will deteriorate for the other problems, as similar satellite readings might have to predict higher hour–ahead PV energy values in the morning but lower values in the afternoon. Because of this, we will build for all these problems three submodels adjusted to different day hours which roughly approximate morning (M), noon (N) and evening (E). For the models that relate radiances at hour H with PV energy at the same hour, we will use the following hour subsets and notation:

- Model m0M for hours [5, 6, 7, 8, 9, 10];
- Model m0N for hours [11, 12, 13, 14];
- Model m0E for hours [15, 16, 17, 18, 19, 20].

For the other models relating $H - 1$, $H - 2$ and $H - 3$ radiances with hour H PV energy, models m1N, m2N, m3N and m1E, m2E, m3E, will use the same hour subsets of models m0N and m0E respectively, while model m1M will use hours [6, 7, 8, 9, 10], m2M hours [7, 8, 9, 10] and m3M hours [8, 9, 10].

We normalize each feature to 0 mean and 1 standard deviation. While this is not actually needed for Lasso, as it is a linear model, it is crucial (and customarily done) for Gaussian SVR, to control for large value effects over the Gaussian kernel. The hyper–parameters for the twelve models we have used are in Table 1; notice the fairly big C and small ϵ values, that suggest a tight, small training error model. Also, the γ values are close to SVR's default $1/d = 7.37 \times 10^{-5}$.

Table 3. Average SVR test errors of m1 models per hour and month.

Month	Hour														
	6	7	8	9	10	11	12	13	14	15	16	17	18	19	20
January	0.00	0.43	0.39	1.61	2.68	3.49	3.60	3.64	3.71	2.82	2.34	0.96	0.19	0.00	0.00
February	0.01	0.34	1.01	2.80	3.29	4.78	3.96	4.25	5.80	4.65	4.35	2.16	0.76	0.01	0.00
March	0.24	0.57	2.56	3.25	3.43	4.04	4.02	4.19	4.08	4.03	2.96	2.13	1.59	0.23	0.00
April	0.47	1.57	2.06	3.09	3.98	3.81	3.90	4.05	3.35	5.03	3.87	2.71	2.13	1.11	0.01
May	0.85	2.53	2.52	2.52	3.14	2.62	2.50	2.30	2.41	2.30	2.54	2.65	2.46	2.20	0.37
June	1.33	3.59	3.46	2.70	1.80	1.79	1.48	1.36	1.66	2.29	3.00	4.31	4.47	3.07	0.67
July	0.96	2.07	1.88	2.19	1.72	2.05	1.21	1.23	1.12	1.68	2.10	3.03	3.45	3.05	0.82
August	0.61	2.40	3.01	2.69	2.04	2.17	2.09	2.18	2.55	2.57	4.05	4.71	4.23	1.99	0.48
September	0.24	1.49	2.04	2.78	2.78	2.71	2.70	2.51	2.85	3.31	3.51	3.12	2.17	0.45	0.00
October	0.44	0.56	2.51	3.69	4.45	4.31	4.40	5.15	5.78	4.20	3.30	2.01	0.64	0.00	0.00
November	0.01	0.40	0.86	2.56	3.55	4.66	4.76	4.87	4.86	3.04	2.54	0.80	0.04	0.00	0.00
December	0.00	0.41	0.47	2.25	3.00	7.26	5.20	5.29	6.05	2.99	1.79	0.73	0.00	0.00	0.00
Average	0.43	1.36	1.90	2.68	2.99	3.64	3.32	3.42	3.68	3.24	3.03	2.44	1.84	1.01	0.20

To achieve a more homogeneous comparison across problems, we will report average errors for UTC hours between 8 and 20 (i.e., those considered for the $H - 3$ problem). These are given for Lasso and SVR in Table 2; for each month we compute errors only on its daylight hours to discard the trivial prediction of no energy at night hours. As it can be seen, the best results for both approaches are achieved for the more stable mid year months and the $H - 1$ problem. While at first sight one should expect similar results for the H problem (that we took as a control problem, as it has no practical interest), this is not the case, as its errors are closer to those of the $H - 2$ problem. This is most likely due to the fact that before noon radiance at hour H overestimates energy production ending at hour H, while underestimates it after noon; similarly, radiance at hour H overestimates radiance at hour ending at $H + 2$. (PV energy readings for Peninsular Spain are available within the 10 min following each hour.) Worst errors are obtained, as expected, for the $H - 3$ problem, but they are too heavily influenced by large errors of the submodel m3M, that has to predict PV readings late in the morning from radiances near sunrise, something that is obviously very difficult for winter days, with very small irradiances at dawn hours and where the model prediction essentially reduces to the bias term. In general, the more powerful SVR models improve on the Lasso predictions. Table 3 gives the average test errors per hour and month of the m1 SVR models (the zero errors in early and late December and January correspond to night hours).

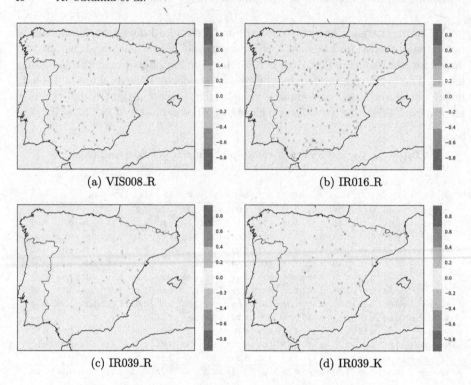

(a) VIS008_R   (b) IR016_R

(c) IR039_R   (d) IR039_K

Fig. 3. Lasso coeffcients for the m1N submodel.

As mentioned, both ML approaches lend themselves to further interpretation. In Fig. 3 we show the Lasso coefficients for each channel used by the m1N submodel (i.e., noon hours in the $H - 1$ problem) as a heat map, where dark blue means large negative coefficients and dark red large positive ones. Since we normalize each feature to 0 mean and 1 standard deviation, the linear coefficient values give a rough idea of the variable influence. For better visualization all figures have the same colormap scale. First, as expected, Lasso imposes sparsity on the models, as clearly non-zero coefficients only appear at few grid points. Notice also that there are more non-zero coefficients for IR016 radiances which also attain the largest scale values; the situation is more or less the opposite for the IR039 ones. In any case, this should be studied further. For instance there are quite a few coefficients for grid points in northern Spain, which has a much smaller number of PV plants; moreover compensating effects among variables seem to appear as there are quite a few negative coefficients, some of them with large values and often quite close to the locations of the positive coefficients.

Recall that we can interpret SVR coefficients from a time perspective looking at how many patterns at the possible day-hour pairs are taken as Support Vectors (SV) by the different m1, m2 and m3 models. However no sample sparsity is achieved in this way; for instance, out of a maximum of 5,475 possible SVs for the m1 models, a total of 4,977 have been selected and similar values appear for the m2 and m3 models. This is opposite to the Lasso models where, for instance, the m1N model has 874 non-zero coefficients while the number of Lasso variables is 13,564, i.e., 4 variables × 3,391 grid points. Note, however, that the data matrix has a maximum rank of 1,460, i.e., 4 noon hours × 365 days, which is much closer to the number of non-zero Lasso coefficients.

5   Discussion and Conclusions

In this work we have studied the use of Meteosat irradiances to nowcast the PV energy production of Penisular Spain from a Machine Learning point of view. More precisely, we have applied two well known ML methods, Lasso and Support Vector Regression, to predict energy production at hour H from satellite readings at hours $H, H-1, H-2$ and $H-3$. The best forecasts for both models are those from readings at hour $H-1$, followed more or less evenly by those at hours H and $H-2$ and with those at hour $H-3$ being the worst; moreover, SVR forecasts were better than Lasso ones. These results were to be expected except, perhaps, the poorer results of the H readings although, as mentioned, being the readings irradiances at the hour, they overshoot energies in the first half of the day and undershoot them in the second half. More important may be the concrete error values, although they are difficult to compare, as PV energy is highly dependent on geography, and errors for Spain simply will not be comparable with those given in the literature for, say, Germany; the concrete temporal periods studied may also have a noticeable influence.

In any case, there are several ways to improve on the results presented. To begin with, a better choice than the single readings at hour H is given by the averages of the irradiance readings at the four 15' periods that end at that hour (these readings are also provided by EUMETSAT). We can also adjust more closely the Meteosat grid points to the Spanish areas with substantial PV capacity and another clear improvement is to better adapt the training data to the forecast period to take account of seasonal effects. For instance, in [2] the forecast for month M of year Y was obtained using as training data months $M-1, M-2$ of the same year and months $M, M+1$ and $M+2$ of the previous year. This results in month-adjusted models and should give better results than our full year approach, which should have some difficulties balancing the very different solar regimes of, say, summer and winter. Finally, methods such as Heliosat give modelling alternatives and other sources of information can be put to play. For instance, one can contemplate energy nowcasting as a correction of previous energy forecasts, for which day ahead NWP irradiance forecasts and past PV energy readings can be put to good use, as shown in [2]. We are currently pursuing these and other related ideas.

Acknowledgments. With partial support from Spain's grants TIN2013-42351-P (MINECO), the UAM-ADIC Chair for Data Science and Machine Learning and S2013/ICE-2845 CASI-CAM-CM (Comunidad de Madrid). The first author is kindly supported by the UAM-ADIC Chair for Data Science and Machine Learning and the second author by the FPU-MEC grant AP-2012-5163. We gratefully acknowledge the use of the facilities of Centro de Computación Científica (CCC) at UAM and thank Red Eléctrica de España for kindly supplying PV energy data.

References

1. Antonanzas, J., Osorio, N., Escobar, R., Urraca, R., de Pison, F.M., Antonanzas-Torres, F.: Review of photovoltaic power forecasting. Sol. Energy **136**, 78–111 (2016)
2. Fernández-Pascual, Á., Gala, Y., Dorronsoro, J.R.: Machine learning prediction of large area photovoltaic energy production. In: ECML PKDD Workshop on Data Analytics for Renewable Energy Integration DARE 2014, pp. 38–53 (2014)
3. Fonseca, J.G.S., Oozeki, T., Takashima, T., Koshimizu, G., Uchida, Y., Ogimoto, K.: Photovoltaic power production forecasts with support vector regression: a study on the forecast horizon. In: 2011 37th IEEE on Photovoltaic Specialists Conference (PVSC), pp. 002579–002583 (2011)
4. Hammer, A., Heinemann, D., Hoyer, C., Kuhlemann, R., Lorenz, E., Müller, R., Beyer, H.G.: Solar energy assessment using remote sensing technologies. Remote Sens. Environ. **86**(3), 423–432 (2003)
5. Hastie, T., Tibshirani, R., Friedman, J.: The Elements of Statistical Learning. Springer, Heidelberg (2009)
6. Inman, R., Pedro, H., Coimbra, C.: Solar forecasting methods for renewable energy integration. Prog. Energy Combust. Sci. **39**(6), 533–576 (2013)
7. Kühnert, J., Lorenz, E., Heinemann, D.: Satellite-based irradiance and power forecasting for the German energy market. In: Kleissl, J. (ed.) Solar Energy Forecasting and Resource Assessment, pp. 267–297. Academic Press (2013)
8. Kleissl, J.: Solar Energy Forecasting and Resource Assessment. Academic Press, Cambridge (2013)
9. Lorenz, E., Kühnert, J., Wolff, B., Hammer, A., Kramer, O., Heinemann, D.: PV power predictions on different spatial and temporal scales integrating PV measurements, satellite data and numerical weather predictions. In: Proceedings of the 29-th European Photovoltaic Solar Energy Conference and Exhibition (EU PVSEC), pp. 22–26 (2014)
10. Marquez, R., Coimbra, C.F.: Intra-hour DNI forecasting based on cloud tracking image analysis. Sol. Energy **91**, 327–336 (2013)
11. Mohammed, A.A., Yaqub, W., Aung, Z.: Probabilistic forecasting of solar power: an ensemble learning approach. In: Neves-Silva, R., Jain, L.C., Howlett, R.J. (eds.) Intelligent Decision Technologies. SIST, vol. 39, pp. 449–458. Springer, Heidelberg (2015). doi:10.1007/978-3-319-19857-6_38
12. Pedro, H.T., Coimbra, C.F.: Assessment of forecasting techniques for solar power production with no exogenous inputs. Sol. Energy **86**(7), 2017–2028 (2012)
13. Rana, M., Koprinska, I., Agelidis, V.G.: 2D-interval forecasts for solar power production. Sol. Energy **122**, 191–203 (2015)
14. Schölkopf, B., Smola, A.: Learning with Kernels: Support Vector Machines, Regularization, Optimization, and Beyond. MIT Press, Cambridge (2001)
15. Wolff, B., Kühnert, J., Lorenz, E., Kramer, O., Heinemann, D.: Comparing support vector regression for PV power forecasting to a physical modeling approach using measurement, numerical weather prediction, and cloud motion data. Sol. Energy **135**, 197–208 (2016)

Approximate Probabilistic Power Flow

Carlos D. Zuluaga$^{(\boxtimes)}$ and Mauricio A. Álvarez

Department of Electrical Engineering, Faculty of Engineering,
Universidad Tecnológica de Pereira, Pereira, Colombia 660003
{cardazu,malvarez}@utp.edu.co

Abstract. Power flow analysis is a necessary tool for operating and planning Power systems. This tool uses a deterministic approach for obtaining the steady state of the system for a specified set of power generation, loads, and network conditions. However this deterministic methodology does not take into account the uncertainty in the power systems, for example the variability in power generation, variation in the demand, changes in network configuration. The probabilistic power flow (PPF) study has been used as an useful tool to consider the system uncertainties in power systems. In this paper, we propose another alternative for solving the PPF problem. This paper shows a formulation of the PPF problem under a Bayesian inference perspective and also presents an approximate Bayesian inference method as a suitable solution of a PPF problem. The proposed method assumes a solution drew from a prior distribution, then it obtains simulated data (active and reactive power injected) using power flow equations and finally compares the observed data and simulated data for accepting the solution or rejecting these variables. This overall procedure is known as Approximate Bayesian Computation (ABC). An experimental comparison between the proposed methodology and traditional Monte Carlo simulation is also shown. The proposed methods have been applied on a 6 bus test system and 39 bus test system modified to include a wind farm. Results show that the proposed methodology based on ABC is another alternative for solving the probabilistic power flow problem; similarly this approximate method take less computation time for obtaining the probabilistic solution with respect to the state-of-the-art methodologies.

1 Introduction

A power flow study is an essential tool for the analysis, operation and planning of a power system (PS). When the power generation, loads and the network configuration are specified, it is possible to compute the steady-state of the PS. It is common to use constant parameters to solve the power flow problem (Aien et al. 2014). However, in a real PS, the load demand, network configuration and power supplies are not fixed, due to uncertainty of customer load demand, uncertainty of equipment failure and operation, and uncertainty of weather. Therefore, a deterministic power flow study does not compute correctly the state of the system (Soleimanpour and Mohammadi 2013; Sansavini et al. 2014).

© Springer International Publishing AG 2017
W.L. Woon et al. (Eds.): DARE 2016, LNAI 10097, pp. 43–53, 2017.
DOI: 10.1007/978-3-319-50947-1_5

A probabilistic power flow (PPF) analysis assumes that specific variables in the system can be treated as random variables with particular probability distributions, where the PPF goal is to obtain probability distributions over voltages, angles and power flows between lines. The solution methods for PPF problem can be classified into two main categories: simulation-based methods or analytical methods (Aien et al. 2014). In the first category, the most widely used method is Monte Carlo simulation (MCS). This method uses repeated deterministic solutions when the input random variables (loads, power generation and voltages at PV nodes) follow an associated distribution function that accounts for the uncertainty. The second category includes approaches based on probabilistic analysis intervals (Briceno et al. 2012), cumulant methods (Le et al. 2013) and point estimation methods (Su 2005). These methods are computationally more effective than methods based on simulation, however they require mathematical assumptions or approximations for feasible solutions (Soleimanpour and Mohammadi 2013). Hence, the analytical methods may offer less accurate solutions than approaches based on simulation methods (Soleimanpour and Mohammadi 2013).

In this paper, we use simulation-based methods under a Bayesian inference perspective. We assume that the state variables (angles at PQ and PV nodes, and voltages at PQ nodes) have a prior distribution, and we want to obtain the posterior distribution for these variables. From the point of view of Bayesian inference, it is necessary to specify a prior distribution of state variables and a likelihood function for the PPF problem, and then using the Bayes theorem, the posterior distribution for these variables can be computed.

Different approaches have used Bayesian inference applied to power systems. In Carmona-Delgado et al. (2015), the authors model the input random variables as multivariate Gaussian mixture distributions and use the expectation maximization algorithm for obtaining samples from these mixture distributions and then, they use these random samples configuration and deterministic optimization for calculating the probability distributions of state variables. Another study uses a kernel density estimation for inferring the probability distribution of wind speed and explores the impacts of high dimensional dependences of wind speed among wind farms on the PS (Cao and Yan 2017). However, they do not use a likelihood function of the powers injected to the PS given the state variables. Since the likelihood function for the PPF problem has not been defined, we need either to propose a likelihood function or to use likelihood-free Monte Carlo approaches. In this paper, we use likelihood-free inference. Within of likelihood-free methods, Approximate Bayesian Computation (ABC) can be employed to infer posterior distributions without having to evaluate likelihood functions (Pritchard et al. 1999; Wilkinson 2013).

We evaluate the performance of two likelihood-free algorithms, namely, ABC and ABC MCMC. ABC is based on rejection sampling (Wilkinson 2013). We also use ABC MCMC which it combines the ABC approach and the Markov Chain Monte Carlo (MCMC) method (Marjoram et al. 2003). We apply the ABC methods and MCS over 2 test systems: the IEEE 6-bus test system and a 39-bus test system modified.

2 Power Flow Analysis

According to Su (2005), the power flow analysis can be expressed by two sets of nonlinear equations. Given the network configuration, the power flow equations can be written as follows,

$$\mathbf{b} = \mathbf{g}(\mathbf{x}), \tag{1}$$

$$\mathbf{z} = \mathbf{h}(\mathbf{x}), \tag{2}$$

where $\mathbf{g}$ and $\mathbf{h}$ are nonlinear power flow equations; $\mathbf{x}$ is a state variable vector that contains the angles at the PQ and PV nodes, and the voltages at the PQ nodes; $\mathbf{b}$ is a vector with entries given by the net active power and the net reactive power injected, which are known. $\mathbf{z}$ is a vector which the elements are the power flows through lines. To solve the power flow problem (obtain $\mathbf{x}$ using Eq. (1)), it is common to use the Gauss-Seidel method or the Newton-Raphson method iteratively (Wood and Wollenberg 1996), then it is possible to compute the power flow employing Eq. (2).

We propose to use Eq. (1) for obtaining an approximation of the probability distribution over the voltages and angles, when the input data is modeled by combinations of input random variables. The goal of our approach is to use a Bayesian perspective for inferring a probability distribution over $\mathbf{x}$ given observed data or powers injected considering uncertainty. In our context, and using Bayes theorem,

$$p(\mathbf{x}|\mathcal{D}) = \frac{p(\mathcal{D}|\mathbf{x})\,p(\mathbf{x})}{p(\mathcal{D})}, \tag{3}$$

where $p(\mathbf{x})$ is the prior distribution (prior) for $\mathbf{x}$, $p(\mathcal{D}|\mathbf{x})$ is the likelihood function, and $p(\mathbf{x}|\mathcal{D})$ is the posterior probability distribution (posterior) of the state variable $\mathbf{x}$ given the observed data $\mathcal{D}$. The posterior quantifies the knowledge about the unknown variables, and evaluate the uncertainty in $\mathbf{x}$ after observing $\mathcal{D}$ (Murphy 2012). The term $p(\mathcal{D})$ is a normalization constant, and it is given by (Murphy 2012),

$$p(\mathcal{D}) = \int p(\mathcal{D}|\mathbf{x})\,p(\mathbf{x})\,d\mathbf{x}. \tag{4}$$

From Eq. (3), posterior, and Eq. (4), evidence, depend on a likelihood function, and as we mentioned above, the likelihood function for PPF problem has not been defined, therefore we can not compute an analytical solution in a closed-form for the posterior and the evidence. For this reason, we need either to construct a likelihood function or to work with likelihood-free methods for obtaining the posterior. In this paper, we employ likelihood-free inference or Approximate Bayesian Computation methods, that consists in generating samples (powers injected) through a simulator. For the PPF problem, this simulator is given by expression $\mathbf{g}(\mathbf{x})$ in Eq. (1).

3   Materials and Methods

IEEE 6-bus Test System. For our experiments, we use the IEEE 6-bus test system shown in Su (2005) (Fig. 2). From Su (2005), we employ the same nodal data, and consider the transmission line parameters without uncertainty. Specifically, we model active and reactive power demands through Gaussian distributions. For active power generation at buses 2 and 3, we use Gaussian distributions. For the voltage at PV nodes, that is, voltages at buses 2 and 3, we also employ Gaussian distributions.

IEEE 39-bus Test System Modified. As part of our experiments, we also use the IEEE 39-bus test system.[1] In this case study, we modeled all input random variables (real and reactive demanded power at PQ nodes; voltage magnitudes and real power injected at PV nodes) from Gaussian distributions; and add a wind farm at bus 32, similar to Soleimanpour and Mohammadi (2013), where the output power from one wind turbine is given by

$$
P_w\left(v_w\right) = \begin{cases} 0 & v_w \leq v_{cin}, \\ 0.5\rho A_w C_p v_w^3 & v_{cin} < v_w \leq v_r, \\ P_r & v_{cout} < v_w, \end{cases} \tag{5}
$$

where ρ is the air density; $A_w = \pi R^2$ is the area of the wind turbine rotor; R is the radius of the rotor; C_p is a coefficient of power; v_{cin} is the cut-in wind speed, at which the wind turbine generator starts generating power (Sansavini et al. 2014); v_{cout} is cut-out wind speed, at which the wind turbine generator is shut down for safety reasons (Sansavini et al. 2014); v_r is the nominal rotational speed; P_r is the nominal wind power; and v_w is the wind speed that is assumed to follow a Weibull distribution (Soleimanpour and Mohammadi 2013),

$$
f\left(v_w \,|\, a, b\right) = \frac{b}{a}\left(\frac{v_w}{a}\right)^{b-1} e^{-\left(\frac{v_w}{a}\right)^b}, \tag{6}
$$

where a and b are the scale and shape parameters, respectively. For our experiments, we use $a = 15$ and $b = 2.5$. For the parameters in Eq. (5), we use $v_{cin} = 3\,\mathrm{m/s}$, $v_{cout} = 25\,\mathrm{m/s}$, $v_r = 10.28\,\mathrm{m/s}$, $C_p = 0.473$ and $R = 45\,\mathrm{m}$, We adopt similar parameters to the ones used by Soleimanpour and Mohammadi (2013).

In this paper, we also model the wind farm output power as a Gaussian random variable, similar to Sansavini et al. (2014), the output power can be modeled as: $P_{wt}\left(v_w\right) = P_w\left(v_w\right) + \omega$, where $P_{wt}\left(v_w\right)$ is the wind farm output power; $P_w\left(v_w\right)$ is the deterministic output power expressed by Eq. (5); and ω is a white Gaussian noise, that is, ω is modeled using a Gaussian distribution with zero mean value and variance $\sigma_\omega^2 = 0.000001$. The output power from the wind farm is shown in Fig. 1.

---

[1] This systems is available at http://www.pserc.cornell.edu/matpower/.

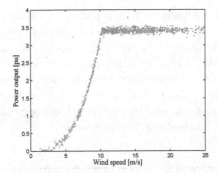

Fig. 1. Output power from wind generator connected at node 32.

Approximate Bayesian Computation Methods. In this paper, we use likelihood-free Monte Carlo approaches, specifically Approximate Bayesian Computation (ABC) methods, as an alternative for solving the PPF problem. Given a prior distribution $p(\mathbf{x})$, the goal in ABC is to approximate the posterior distribution using the following model,

$$p(\mathbf{x}|\mathcal{D}) \propto g(\mathcal{D}|\mathbf{x})p(\mathbf{x}), \qquad (7)$$

where $g(\cdot)$ is a function that depends on the model. These likelihood-free algorithms replace the computation of the likelihood with a comparison between summary statistics of the observed data (active and reactive power injected) and summary statistics from simulated data. The most basic ABC algorithm is based on rejection sampling, and it is given by (Wilkinson 2013).

Algorithm 1. ABC rejection

Draw $\mathbf{x}$ from $p(\mathbf{x})$
Simulate $\mathcal{D}'$ using $g(\cdot|\mathbf{x})$
Accept $\mathbf{x}$ if $\rho(s(\mathcal{D}), s(\mathcal{D}')) \leq \epsilon$

---

where $\rho(\cdot)$ is a distance measure; ϵ is a tolerance that determines the accuracy of the algorithm; $s(\mathcal{D})$ are the summary statistics from observed data, and $s(\mathcal{D}')$ are summary statistics from simulated data. For our study, $\mathcal{D}$ are injections of reactive and active power. It is important to mention that the empirical distribution over accepted samples for $\mathbf{x}$ is an approximation to the true posterior distribution, the approximation can be expressed through $p(\mathbf{x}|\rho(s(\mathcal{D}), s(\mathcal{D}')) \leq \epsilon)$. If $\epsilon = 0$, the samples that we draw will come from the true posterior, however the algorithm would need to perform more simulations for accepting any sample (Wilkinson 2013). Since choosing $\epsilon = 0$ would be prohibitively expensive, the value of ϵ that we choose will affect the quality of the approximation. From Algorithm 1, another issue to address how to choose a suitable distance measure or the summary statistics. See Wilkinson (2013) for details. To avoid some of these problems, several ABC approaches have been proposed, like ABC MCMC (Marjoram et al. 2003), that is described below.

ABC MCMC. In the Algorithm 1, it is possible to have low acceptance rates when the prior distribution is not close to the posterior distribution. An alternative solution for this problem is provided by a MCMC approach (Marjoram et al. 2003). ABC MCMC proposes a new parameter value using a proposal distribution $(q(\cdot))$. This distribution must be chosen so that we can easily evaluate it, and generate samples from it. For more information about the proposal distribution and its parameters, see Murphy (2012), chapter 24. Algorithm 2 shows how the ABC MCMC method can be implemented.

Algorithm 2. ABC MCMC

Initialize $\mathbf{x}_0$
for $i = 1, \ldots, N_s$ do
    Draw $\mathbf{x}^*$ from $q(\mathbf{x} \mid \mathbf{x}_i)$
    Simulate $\mathcal{D}'$ using $g(\cdot \mid \mathbf{x}^*)$
    if $\rho(s(\mathcal{D}), s(\mathcal{D}')) \leq \epsilon$
    Accept $\mathbf{x}_{i+1} = \mathbf{x}^*$ with probability
    $\alpha = \min\left(1, \frac{p(\mathbf{x}^*)q(\mathbf{x}_i \mid \mathbf{x}^*)}{p(\mathbf{x}_i)q(\mathbf{x}^* \mid \mathbf{x}_i)}\right)$
    Otherwise $\mathbf{x}_{i+1} = \mathbf{x}_i$

Algorithm 2 also obtains samples from an approximated posterior over $\mathbf{x}$ (Toni et al. 2009). However, according to Toni et al. (2009), ABC MCMC may get stuck in regions of low probability for the accepted samples. For avoiding this, we propose to use a multivariate Gaussian distribution as proposal distribution, where its mean value can be computed using power system parameters. We take $\mathbf{x}_0$ as an all-ones vector for voltages and the solution of DC power flow algorithm for angles.

3.1   Classic Probabilistic Power Flow Analysis

Classic probabilistic power flow analysis or Monte Carlo simulation (MCS) is based on repetitive solutions to the DLF problem (i.e. $\mathbf{g}(\mathbf{x}) - \mathbf{b} = \mathbf{0}$ from Eq. (1)) using a set of random vectors drawn from input random variables. In this paper, we use the Newton Raphson method as deterministic method of solution, and use MATPOWER to implement MCS.[2]

3.2   Validation

For validation purposes, we compute the relative error (RE) between the value obtained using power flow analysis without considering uncertainty over variables and the mean value for each variable obtained using the simulation methods. The RE is given by

---

[2] This package is available at http://www.pserc.cornell.edu/matpower/.

$$RE = \left\| \frac{x_{wu} - x_{mp}}{x_{wu}} \right\|, \tag{8}$$

where x_{wu} is the original value without considering uncertainty over variables, and x_{mp} is the estimated posterior mean for that variable obtained through each simulation method. This error is only for the IEEE 6-bus test system. For the IEEE-39 mentioned above, we compute relative errors with respect to the values calculated using MCS, that is, we employ (Su 2005),

$$\varepsilon_x^\mu = \left| \frac{\mu_x - \mu_x^*}{\mu_x} \right|, \quad \varepsilon_x^\sigma = \left| \frac{\sigma_x - \sigma_x^*}{\sigma_x} \right|, \tag{9}$$

where ε_x^μ and ε_x^σ are relative error for the mean and standard deviation values; μ_x and σ_x are the mean and standard deviation obtained using MCS; μ_x^* and σ_x^* are the mean and standard deviation computed with the ABC methods exposed above.

3.3   Procedure

For the IEEE 6-bus test system, we model the input random variables as Gaussian distributions. The mean values and standard deviations are assumed as mentioned in Su (2005). We also use Gaussian distributions for the voltages at nodes 2 and 3, which are PV buses; mean values are assumed as in Su (2005); and the standard deviations values are set as $\sigma_v = 0.000001$, since Su (2005) did not model these variables as random variables.

For the IEEE 39-bus test system, the input random variables are modeled by Gaussian distributions and a wind farm at bus 32 was added, which includes 2134 wind turbines and the output power from a wind farm was shown in Fig. 1.

From the probability distributions characterizing the above random variables, we generate 1000 samples for each variable for all test systems. We then apply the MCS for each sample configuration. Using this set of samples, we compute $\mathbf{b}_i$ (see Eq. (1)), which is used as observed data for the ABC methods. We then apply ABC and ABC MCMC.

For ABC and ABC MCMC, we use $\epsilon = (0.04, 0.3)$ in the IEEE 6-bus and 39-bus test system, respectively. For both methods, we use a multivariate Gaussian distribution as prior distribution for the voltages, where its mean vector is an all-ones vector, and the covariance matrix is diagonal with parameters σ_V^2. We use the DC power flow algorithm (Wood and Wollenberg 1996), and employ an uniform distribution for each angle as prior distribution for them: we first apply the DC power flow algorithm to the system, from this solution, an uniform distribution is defined ($\theta_i \sim \mathcal{U}(a_i, b_i)$) for each angle. The parameters a_i and b_i will be computed as $a_i = \theta_i^{DC} - \Delta\theta$ and $b_i = \theta_i^{DC} + \Delta\theta$. $\Delta\theta$ quantifies the error arising from the solution obtained by DC power flow algorithm with respect to AC power flow solution (Eq. (1)). We choose $\Delta\theta$ equal to 0.07.

For ABC and ABC MCMC, the simulation function was set as $\mathbf{g}(\mathbf{x})$ (see Eq. (1)). The distance measure that we use is given by

$$\rho(s(\mathcal{D}), s(\mathcal{D}')) = \sqrt{\frac{1}{M}(s(\mathbf{b}) - s(\mathbf{g}(\mathbf{x})))^\top (s(\mathbf{b}) - s(\mathbf{g}(\mathbf{x})))}, \tag{10}$$

where M is the number of injected powers (active and reactive powers). For choosing s, we use a surrogate modeling between $\mathbf{b}$ and $\mathbf{x}$, specifically, we use a feedforward neural network for learning the summary statistics using the approach shown in Fearnhead and Prangle (2012). We use 10 neurons in the hidden layer, employ $\mathbf{b}$ and the mean of solutions obtained by MCS as training data, for both systems.

4   Results and Discussion

In this section, we present a comparison among ABC, ABC MCMC and MCS for solving the PPF problem using the IEEE 6-bus test system and IEEE 39-bus test system mentioned in Sect. 3.

4.1   Results from IEEE 6-bus Test System

ABC, ABC MCMC and MCS were applied over the IEEE 6-bus test system, where the goal is to obtain the posterior for $\mathbf{x} = \begin{bmatrix} \theta_2 \ \theta_3 \ \theta_4 \ \theta_5 \ \theta_6 \ V_4 \ V_5 \ V_6 \end{bmatrix}^\top$, having 1000 samples from the input random variables, i.e. with 1000 different $\mathbf{b}_i$ or $\mathcal{D}_i$ vectors. In Fig. 2, we show the posterior for θ_6, V_6, active and reactive power flow between nodes 5 and 6 (P_{56} and Q_{56}), respectively.

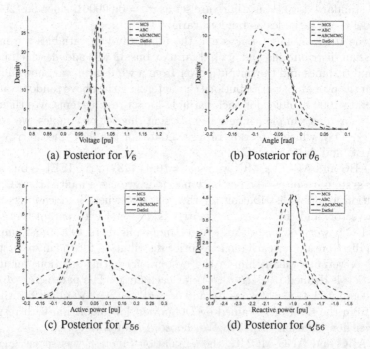

(a) Posterior for V_6

(b) Posterior for θ_6

(c) Posterior for P_{56}

(d) Posterior for Q_{56}

Fig. 2. Posteriors for V_6, θ_6, P_{56} and Q_{56} using ABC ABC MCMC and MCS. The dashed red, blue and black lines are the responses obtained by MCS, ABC and ABC MCMC. The blue vertical solid line is the deterministic solution. (Color figure online)

Notice that the means of the posteriors for the different variables are close to the deterministic solution. The posteriors obtained by ABC MCMC are close to the posteriors using MCS. Since all input random variables are modeled by Gaussian distributions, we can compare the deterministic solution (see the vertical lines of the above figures) of the system and estimated posterior mean using the methods exposed when considering the relative errors (RE) shown in Eq. (8). In Table 1, we show Relative errors for voltages (RE_v) and angles (RE_θ) obtained by MCS, ABC and ABC MCMC. We also present the computation times (CompTime) took by MCS, ABC and ABC MCMC for solving PPF problem using the 6-bus test system. All simulations have been performed on an Intel Core i7 PC with a 2.1 GHz processor. From Table 1, notice that MCS obtained the least RE_v, however this method took the highest CompTime for solving this PPF problem. We note that ABC MCMC took the least CompTime for analyzing this system, and obtained the least RE_θ. Notice that ABC is the least efficient method.

Table 1. RE for voltages (RE_v) and angles (RE_θ) obtained by each method. CompTime is the computation time took by all methods for solving PPF problem using the 6-bus system.

Method	$RE_v[\%]$	$RE_\theta[\%]$	CompTime $[s]$
MCS	**0.0558**	1.4478	21.902
ABC	1.2179	16.337	38.173
ABC MCMC	0.3147	**0.8544**	**17.347**

4.2   Results from IEEE 39-bus Test System

After showing the results using the 6-bus system, we proceed to present the comparison of results obtained by ABC and ABC MCMC with respect to the results employing MCS, when the dimension of the unknown variables is increased.

We use the 39-bus test system, with 67 variables: 38 angles and 29 voltages. In Fig. 3, we present the posterior of θ_{15}, V_{15} and P_{39-9} using all methods exposed. Notice that in this case the ABC does not estimate correctly the posterior of θ_{15}, V_{15} and P_{39-9}. In contrast, the posteriors obtained by ABC MCMC are close to the posteriors using MCS. Since we do not have a ground-truth solution, we use estimated posterior mean and standard deviation (for angles and voltages) of MCS as references values. Using the Eq. (9) and the mean angle values, ABC obtained $\varepsilon_\theta^\mu = 295.12\%$ and ABC MCMC presents $\varepsilon_\theta^\mu = 67.912\%$, and for standard deviation, ABC computed $\varepsilon_\theta^\sigma = 17.345\%$ and ABC MCMC presents $\varepsilon_\theta^\sigma = 18.835\%$. For voltages, ABC obtained a $\varepsilon_V^\mu = 2.5157\%$ and a $\varepsilon_\theta^\sigma = 499.83\%$, ABC MCMC computed a $\varepsilon_V^\mu = 0.5425\%$ and a $\varepsilon_\theta^\sigma = 8.4428\%$. Notice that ABC MCMC obtained the best results. ABC suffers when the dimension of the unknown variables is increased. Finally, MCS took 24.211s for solving the PPF problem, ABC MCMC employed 18.361s and ABC took 35.895s.

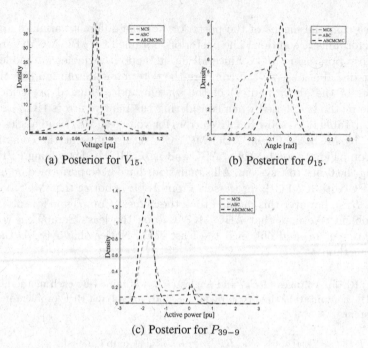

(a) Posterior for V_{15}. (b) Posterior for θ_{15}.

(c) Posterior for P_{39-9}

Fig. 3. Posteriors for θ_{15}, V_{15} and P_{39-9} using ABC ABC MCMC and MCS. The dashed red, blue and black lines are the responses obtained by MCS, ABC and ABC MCMC. (Color figure online)

5   Conclusions

We introduced a new alternative for solving the PPF problem, using approximate Bayesian computation methods. We demonstrated that ABC can work for small system using Gaussian random variables. We also showed that the posteriors of the sate variables obtained by ABC MCMC are close to the results using MCS, similarly this approximate method take less computation time for obtaining the probabilistic solution with respect to MCS.

Acknowledgements. C.D. Zuluaga is being funded by Department of Science, Technology and Innovation, Colciencias. This work was developed within the research project: "Approximate Bayesian Computation applied to probabilistic power flow" financed by Universidad Tecnológica de Pereira, Colombia.

References

Aien, M., Fotuhi-Firuzabad, M., Rashidinejad, M.: Probabilistic optimal power flow in correlated hybrid wind-photovoltaic power systems. IEEE Trans. Smart Grid **5**(1), 130–138 (2014). doi:10.1109/TSG.2013.2293352. ISSN 1949-3053

Briceno, W.C., Caire, R., Hadjsaid, N.: Probabilistic load flow for voltage assessment in radial systems with wind power. Int. J. Electr. Power Energy Syst. **41**(1), 27–33 (2012). ISSN 0142–0615

Cao, J., Yan, Z.: Probabilistic optimal power flow considering dependences of wind speed among wind farms by pair-copula method. Int. J. Electr. Power Energy Syst. **84**, 296–307 (2017). ISSN 0142–0615

Carmona-Delgado, C., Romero-Ramos, E., Riquelme-Santos, J.: Probabilistic load flow with versatile non-Gaussian power injections. Electr. Power Syst. Res. **119**, 266–277 (2015). doi:10.1016/j.epsr.2014.10.006. ISSN 0378–7796

Fearnhead, P., Prangle, D.: Constructing summary statistics for approximate Bayesian computation: semi-automatic approximate Bayesian computation. J. Roy. Stat. Soc.: Ser. B (Stat. Methodol.) **74**(3), 419–474 (2012). ISSN 1467–9868

Le, D.D., Berizzi, A., Bovo, C., Ciapessoni, E., Cirio, D., Pitto, A., Gross, G.: A probabilistic approach to power system security assessment under uncertainty. In: Bulk Power System Dynamics and Control - IX Optimization, Security and Control of the Emerging Power Grid (IREP), 2013 IREP Symposium, pp. 1–7, August 2013. doi:10.1109/IREP.2013.6629411

Marjoram, P., Molitor, J., Plagnol, V., Tavaré, S.: Markov chain Monte Carlo without likelihoods. Proc. Natl. Acad. Sci. U.S.A. **100**(26), 15324–15328 (2003). doi:10.2307/3149004. ISSN 00278424

Murphy, K.P.: Machine Learning: A Probabilistic Perspective (Adaptive Computation and Machine Learning Series). MIT Press, Cambridge (2012). ISBN 0262018020

Pritchard, J.K., Seielstad, M.T., Perez-Lezaun, A., Feldman, M.W.: Population growth of human Y chromosomes: a study of Y chromosome microsatellites. Mol. Biol. Evol. **16**(12), 1791 (1999)

Sansavini, G., Piccinelli, R., Golea, L.R., Zio, E.: A stochastic framework for uncertainty analysis in electric power transmission systems with wind generation. Renew. Energy **64**, 71–81 (2014). doi:10.1016/j.renene.2013.11.002. http://www.sciencedirect.com/science/article/pii/S0960148113005806. ISSN 0960–1481

Soleimanpour, N., Mohammadi, M.: Probabilistic load flow by using nonparametric density estimators. IEEE Trans. Power Syst. **28**(4), 3747–3755 (2013). doi:10.1109/TPWRS.2013.2258409. ISSN 0885–8950

Su, C.: Probabilistic load-flow computation using point estimate method. IEEE Trans. Power Syst. **20**(4), 1843–1851 (2005). doi:10.1109/TPWRS.2005.857921. ISSN 0885–8950

Toni, T., Welch, D., Strelkowa, N., Ipsen, A., Stumpf, M.: Approximate Bayesian computation scheme for parameter inference and model selection in dynamical systems. J. Roy. Soc. Interface **6**, 187–202 (2009). doi:10.1098/rsif.2008.0172

Wilkinson, R.: Approximate Bayesian Computation (ABC) gives exact results under the assumption of model error. Stat. Appl. Genet. Mol. Biol. **12**(2), 129–141 (2013)

Wood, A.J., Wollenberg, B.F.: Power Generation, Operation, and Control. A Wiley-Interscience Publication. Wiley, Hoboken (1996)

Dealing with Uncertainty: An Empirical Study on the Relevance of Renewable Energy Forecasting Methods

Robert Ulbricht[1(✉)], Anna Thoß[2], Hilko Donker[1], Gunter Gräfe[2], and Wolfgang Lehner[3]

[1] Robotron Datenbank-Software GmbH, Dresden, Germany
{robert.ulbricht,hilko.donker}@robotron.de
[2] Hochschule für Technik und Wirtschaft, Dresden, Germany
anna.thoss@htw-dresden.de, graefe@informatik.htw-dresden.de
[3] Database Technology Group, Technische Universität Dresden, Dresden, Germany
wolfgang.lehner@tu-dresden.de

Abstract. The increasing share of fluctuating renewable energy sources on the world-wide energy production leads to a rising public interest in dedicated forecasting methods. As different scientific communities are dedicated to that topic, many solutions are proposed but not all are suited for users from utility companies. We describe an empirical approach to analyze the scientific relevance of renewable energy forecasting methods in literature. Then, we conduct a survey amongst forecasting software providers and users from the energy domain and compare the outcomes of both studies.

Keywords: Renewable energy forecasting · Practical relevance · Machine Learning

1 Introduction

As much as for any industry, forecasting time series is traditionally an important issue in the energy domain. In corporate areas like distribution or pricing, many decisions always had to be made based on uncertain data. Nowadays, the capacity of renewable energy sources like solar and wind power constantly increases world-wide and with this the necessity of conducting pre-dispatch studies is created. Although this is a standard procedure for conventional power plants where the future energy output can be calculated based on the available fossil resources, the fluctuating nature of renewable energy sources causes additional uncertainty about the expected amount of produced energy and therefore challenges the electric balance between power demand and supply. To allow for their better integration into the power grids and energy markets, the development of precise forecasting approaches for the supply side emerged as a relatively young research topic compared to the long history of energy demand forecasting. With this, interesting new problems arise for researchers and are currently treated by

© Springer International Publishing AG 2017
W.L. Woon et al. (Eds.): DARE 2016, LNAI 10097, pp. 54–66, 2017.
DOI: 10.1007/978-3-319-50947-1_6

a multitude of very active communities. Their results are then implemented in commercial software solutions, brought back to the energy industry and, if successful, adopted by the market. As usual, one problem remains: How to identify the optimal solution propositions for implementation? There are numerous literature reviews available for solar (e.g. Glassley et al. [3] or Yadav and Chandel [12]) and wind power prediction (e.g. Monteiro et al. [8], Colak et al. [2] or Tascikaraoglu and Uzunoglu [10]) and they do give a good overview on the topic or draw attention to recent research trends, but usually few things are said about the overall significance of the discussed approaches within the related research field or in practice. Another alternative is forecasting competitions like *GEF-Com*[1], as they try to bridge the gap between the worlds of science and industry by offering interesting technical challenges to the community. Although such competitions provide well comparable results because all general conditions are pre-defined, the insights published by Hong et al. [4] show the limitations with the simulation of real-world situations, where for example forecasts have to be provided on a rolling basis for intra-day or day-ahead periods. Apparently, there seems to be no serious option beside the classical trial-and-error-approaches. The present work aims at reducing the necessary efforts by providing answers to the following questions: (1) How strong is the current scientific interest in renewable energy forecasting methods, (2) which methods are the preferred research topics, and (3) can the most promising directions be identified? This is then compared to the industry view where we analyze the methods currently implemented in the available software products (4), the quality evaluation criteria commonly used (5) and the users' expectations of such solutions (6).

In this paper, we describe an empirical approach to analyze the current state-of-the-art for forecasting methods used to predict the output of fluctuating renewable sources. The paper consists of 4 sections, the first being this introduction. In Sect. 2, we describe the methodology applied for the empirical review of scientific publications and the obtained results. Section 3 contains a report on a survey conducted among forecasting software providers and users from the energy industry. Finally, we summarize our findings in Sect. 4.

2   Scientific Relevance

In this section, we analyze the scientific relevance of renewable energy forecasting methods based on reviewed literature. We describe the applied methodology before we present and discuss the results. Thereby we differentiate between quantitative and qualitative aspects.

2.1   Methodology

Our first step is to determine a representative population for the analysis. We search the *IEEE*, *ScienceDirect*, *SpringerLink* and *WileyOpenLibrary* online

---

[1] Global Energy Forecasting Competition, http://www.gefcom.org.

databases for relevant articles published during the last decade (2005 to 2015), thus covering the most common publication channels like scientific journals, books and conference proceedings. For the search queries, we combine the keywords *error, forecasting, renewable, wind, solar* and *method* while we exclude *demand* and *production* in order to increase the relevance for our purposes while reducing the total amount of possible hits. This will provide us with an overall idea of the research activity for the target domain across all involved communities.

In the next step, out of the numerous query results obtained, we define a smaller sample data set which is better usable for the in-depth analysis. In order to reduce the risk of getting less relevant or low quality results, we restrict the sample to the most important renewable energy journals according to the *SCImago Journal Rank Indicator* (SJR): Only the highest two quantiles (Q1 and Q2) of journals offering full-text online access are included, which covers the most relevant 50% of all ranked sources. The list of included journals is shown in Table 1. Furthermore, we consider only articles published during the last 6 years (2010 to 2015) thus reflecting the most recent research trends. Finally, all abstracts are manually revised to filter out all topics possibly not relevant for our study. We do explicitly exclude journals originating from other domains, like for example statistical journals. Although offering valuable contributions to the forecasting community, their natural preference for domain-specific approaches (e.g. statistical methods) could lead to biased results.

Table 1. List of relevant journals and number of articles found

Rank	Symbol	Title	SJRQ	SJR	Articles
6	RSER	Renewable and Sustainable Energy Reviews	Q1	3.273	9
7	TSTE	IEEE Transactions on Sustainable Energy	Q1	2.826	8
10	SE	Solar Energy	Q1	2.291	27
11	RE	Renewable Energy	Q1	2.256	27
13	IET	IET Renewable Power Generation	Q1	2.178	1
19	ECM	Energy Conversion and Management	Q1	1.801	12
29	ER	International Journal of Energy Research	Q2	1.106	1
32	EPP	Energy Sources, Part B: Economics, Planning and Policy	Q2	0.856	0
35	WEIA	Journal of Wind Engineering and Industrial Aerodynamics	Q2	0.791	5
42	EPSE	Environmental Progress and Sustainable Energy	Q2	0.629	1
47	RSE	Journal of Renewable and Sustainable Energy	Q2	0.472	2

2.2 Quantitative Analysis

First, we have a look at some quantitative aspects. According to the search criteria previously described in Sect. 2.1, we find a total number of 839 relevant publications. The most important channels in terms of published articles are scientific journals with a share of 47.1%, followed by conference papers with 37.8%, while book chapters are under-represented with 15.1%. Figure 1 demonstrates

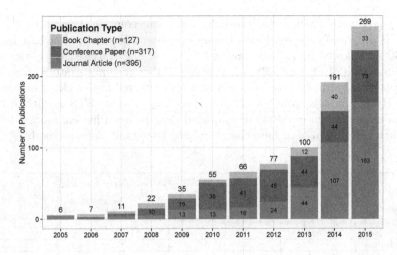

Fig. 1. Evolution of yearly renewable energy forecasting publications

the evolution of the yearly publication frequency for all searched publication types. We observe a steadily increasing number of publications with an average growth of 48.8% p.a. reaching a peak of 269 relevant publications in 2015, the last year considered for the study. This shows a clear trend of the rising and unbroken scientific interest in renewable energy forecasting methods especially during the last 2 years. However, compared to the results we obtain by searching for energy demand (5106 hits) and price forecasting literature (1869 hits) in the same databases, it becomes obvious that renewable energy forecasting still represents only a small subset of less than 11% of the published forecasting activities in the energy domain.

The following results are derived from the reduced sample data set. Out of the 93 filtered articles from the journal search, 83 remain after manual revision. This means that a 10% of the defined population (which still might include less relevant results, as no manual refinement is applied) are represented by the sample. Amongst them, forecasting methods aiming at solar energy are notably over-weighted compared to wind energy with 57.8% vs. 38.6% respectively, and only 3.6% propose methods applicable to both energy sources.

Figure 2 shows the distribution of proposed forecasting methods for solar or wind energy production. According to previous work [11], methods can be classified based on the nature of the underlying modeling technique and the combination type. For example a Neural Network is a Machine Learning technique (ML), while ARIMA is an univariate Stochastic Time Series model (STS) but both of them are statistical models, as they make use of historical observations which is not the case for Physical models. All of them can appear either as a stand-alone method or as part of a combined model. As a consequence, whenever two or more stand-alone methods are combined they cannot be classified unless considering such approaches as a model class on its own, the Hybrid models. Some articles also propose more than one method as stand-alone solution. From a total of 87 methods proposed in that articles, the most common classes are Machine Learning (30%), followed by Hybrid models (29%) and multivariate Stochastic Time Series models (23%), while Physical models (11%) and univariate STS models (5%) play a subordinate role. A similar result is reflected from a closer inspection of the Hybrid models, as again ML (40%) and multivariate STS methods (34%) are the most popular elements of combined models, but Physical models (20%) are much more often used here than they are as a stand-alone method.

Model Class / Combination Type	Physical Model	Statistical Model							Hybrid Model	Σ
		Stochastic Time Series					Similar-Days	Machine Learning		
		Univariat				Multivariat				
		Naive Prediction	Least Squares	Exponential Smoothing	Others					
Stand-alone Model	12	1	1	0	2	20	0	26	25	87
Part of Hybrid Model	10	0	1	1	0	17	1	20	x	50
Σ	22	1	2	1	2	37	1	46	25	137

Fig. 2. Proposed forecasting methods according to model class and combination type

The evolution of the model classes' proportions on the total number of proposed stand-alone models (compare Fig. 3) shows that there are no significant changes in popularity during the observation period. ML and multivariate STS models alternate each other, while only Physical models gain a slightly but constantly increasing share of scientific attention during the last two years covered by this study.

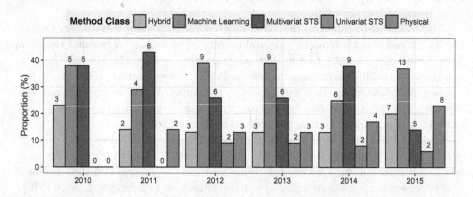

Fig. 3. Temporal evolution of proposed stand-alone model classes. The total yearly publications are shown with discrete values while the bars display the corresponding proportion for each class.

2.3   Qualitative Analysis

After the view on quantitative aspects, the next step to follow in order to determine the most successful forecasting methods is a qualitative analysis of the sample data. This is problematic to the extent that quality determination for the commonly used point forecasts is mostly reduced to the accuracy dimension, sometimes combined with the model's robustness or technical performance. However, the selection of appropriate statistical metrics to measure the accuracy of forecasts is an independent research topic (compare e.g. Chen and Yang [1] or Hyndman and Koehler [5]). For renewable energy forecasts, the foundations of a standardized performance evaluation protocol were defined by Madsen et al. [7] more than a decade ago. As a minimum set of measures, they propose the use of normalized *Mean Bias Error* (MBE), *Mean Absolute Error* (MAE), *Root Mean Square Error* (RMSE) and improvement factors like the *Skill Score* (SS) for comparison between concurring methods and against simple predictors like the *Persistence Model* (PM).

Indeed, as shown in Fig. 4, the most frequently used error measures in the sample data set are RSME (65.1%), MAE (47.0%) and MBE (27.7%). In contrast, only 8 publications (9.6%) make use of all three of them, and this value decreases to 3 (3.6%) when SS is included. The number of simultaneously used accuracy measures reaches from 0 to 8 among all articles with an average value of 2.6 per article. The most frequent combinations are RMSE-MBE and RMSE-MAE with 24.1% and 22.9% respectively, followed by RMSE-MAPE with 14.5%. A 26% make use of normalized source data or error values. We also find that 32% do include a simple benchmark predictor (usually represented by the PM), which all of them outperform. Additionally, there are numerous individual accuracy measures normally seldom found in energy forecasting literature like e.g. *NashSutcliffe Equation*, *Scutter Index* or *Legates-McCabe's Coefficient of Efficiency*. These tailor-made criteria along with all undefined error measures are denoted as 'Others'. Considering the present variety it is not surprising that in the end only one out of the 83 analyzed publications matches all criteria derived from the evaluation protocol described above.

In addition to the problem previously discussed, the qualitative comparability of methods is further delimited by (1) the underlying use case and (2) the operational framework applied in the experiments. Evaluations are conducted on dissimilar data sets, so they do always vary in aspects like measurement quality and completeness, the available input features and history length of the target time series, and individual characteristics like signal-to-noise ratio or aggregation level. The geographical origin of the data also matters, as for example the output from solar energy sources located in northern Europe is less predictable than from sites in the south due to usually less stable weather conditions. Of course, providing access to sensitive real-world data is always problematic but there are alternatives: From the studied articles, 8.4% make use of the publicly available solar or wind data sets offered by the *National Renewable Energy Laboratory* (NREL) [9], which might be a first step towards more homogeneity in use cases. As for the experiments, parameters like forecasting horizon, the ratio of training and evaluation

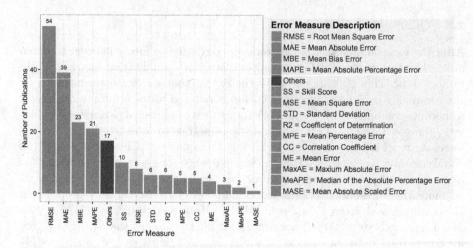

Fig. 4. Variety of applied forecast performance evaluation criteria.

data and any additional pre- or post-processing procedures influence the results. Finally, conclusions can also suffer from the authors' somehow 'personal' point of view. Experimental results may be optimized towards the desired findings by using only very few and specific test cases or compared against too simple baselines, so incredibly low errors or high improvement factors are achieved but will hardly be reproducible in different environments. However, recently published related work proves that such standardization problems still remain unsolved for wind [2] and solar energy [6] alike. Unfortunately, we can not derive reliable statements concerning the performance of the proposed methods for that reason.

3   Practical Relevance

In this section, we analyze the practical relevance of renewable energy forecasting methods for the energy industry. Due to the natural lack of publications in that area, we conduct a brief study of the German energy market to identify possibly relevant target companies and carry out a survey among software providers and users before we compare and discuss the results.

3.1   Methodology

First, we examine the web presences of energy software providers based in Germany and listed in the market reports regularly published by EMW[2]. We verify two simple criteria: (1) The target company offers an energy forecasting software product and (2) the product supports renewable energy supply forecasts. Those companies not providing sufficient information online are contacted directly. Out

---

[2] http://www.emw-online.com.

of 56 revised companies, 6 match both criteria, while 23 match the first criteria only thus giving us a total of 29 potential targets for the survey. We design a questionnaire with 14 questions aiming at obtaining the following information: (1) For which appliance the forecasts are computed, (2) the underlying energy source, (3) the used algorithm classes, (4) additional parameters of interest like forecasting horizon, applied evaluation criteria, pre- or post-processing procedures and (5) some characteristics of the participating company.

For the user questionnaire, we contacted 65 companies from the utility sector, 41 of them located in Germany. This time, we only used 8 questions asking for (1) market role, (2) forecast appliance, (3) underlying energy source, (4) output evaluation criteria and (5) additional parameters like forecasting horizon and weather data sources. We also contacted 13 associations and public organizations belonging to the energy domain and placed posts in energy related news groups to reach as many participants as possible. The surveys were conducted during a period of two months, from August to October 2015.

3.2   Feeback from Software Providers

From the contacted software companies, a total of 19 answers were obtained but only 7 questionnaires are complete and were used. The participants belong to 6 companies, making for a response rate of 21%. Although low in number, if we consider the fact that the focus of our study is a very specialized topic usually treated only by a few corporate experts and may concern sensitive information for potential competitors, having any answer at all is a satisfying result.

According to the answers received, the most common application areas for the commercial software are energy demand and -price forecasting (37.5% each), while only 25% state that the solution is also suitable for energy supply forecasts. No difference is made between the prediction for supply from conventional and renewable energy sources, as all answers state that both can be handled by the software. This indicates that the implemented forecasting algorithms seem to be more general solutions and none of the companies offer a product portfolio explicitly specialized for the renewable energy domain. Furthermore, there is no difference concerning the forecasting horizon as all software products are equally suited for short- (e.g. intra-day or day-ahead) and long-term forecasts. Unfortunately, it remains unclear if the used methods provide a robust result quality during all stages of the target time series and if so, how this is obtained.

Concerning the characteristics of the implemented forecasting methods based on the classification previously introduced in Sect. 2.2, 21 methods are mentioned: With 29% the most common method is the Similar-Days approach, followed by Machine Learning and Stochastic Time Series models with 24% each. Hybrid (14%) and physical models (10%) seem to be of less significance for the software providers. Hereby, no difference is made between stand-alone methods or methods used as part of hybrid models. Beside the model class, some participants provide additional information as they specify *Multi-variate Regression* (5 times), *Neural Networks* (4 times), *k-nearest Neighbors* (2 times) and *Support Vector Machines* (1) as the underlying algorithms of their product. Figure 5 compares

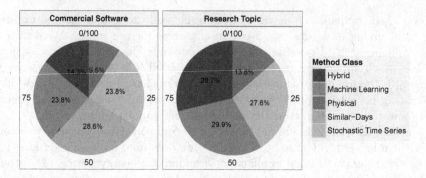

Fig. 5. Comparison of scientific and practical relevance of stand-alone method classes. Showing the proportions of methods implemented in commercial software on the left and corresponding research topics (compare Fig. 2) on the right side.

these results to the values obtained from the scientific literature: While STS and ML models show an almost identical relevance and dominate both areas, it is noted that the practical interest in hybrid and physical models is much inferior. On the other hand, the similar-days method plays a much more significant role in industry but is irrelevant for scientists.

Further, 50% of the participants state that data pre-processing procedures are applied, 40% mention data post-processing procedures and almost half (42%) use both of them. All answers claim that the provided solution can handle seasonal influences. Concerning the applied evaluation criteria, a 25% uses Standard Deviation (SD), while 21% use each RMSE (respective nRMSE) and MAPE and only 13% use MAE. Other mentioned error measures are BIAS, WMAE and U-Theil.

3.3 Feeback from Software Users

For the user questionnaire, we obtain 14 completed and 35 partially completed responses. From the latter 6 could be used as they contain sufficient pieces of information for our purposes, making for a response rate of 21.5% or 30% respectively. Concerning the participant's market roles, 25% of the companies classify as transmission system operators (TSO), each 20% as energy suppliers and -generators and 10% as distribution network operators (DSO), while 15% classify as others (e.g. energy planning &construction, independent energy procurer) and 10% do not classify.

A 42.9% of the participants state that they are already using energy forecasting software, while 38.1% do not. The almost equal distribution of the results indicates a still unsaturated market and thus a promising potential for such solutions. Those using the software tools say that their application focus is the prediction of future energy demand (50%), energy supply (30%) and energy prices (20%). Supply forecasts are mainly computed for renewable energy sources by 73%, and only by 28% for conventional energy production.

Table 2. Importance of positive effects expected from supply forecasts. Values are obtained by dichotomous distribution from 5-level scales.

Factor	Important	Irrelevant	Abstention
Increased supply security	76.5%	5.9%	17.6%
Avoidance of overproduction	64.7%	17.6%	17.6%
Use of smart grid applications	35.3%	41.2%	23.5%
Improved demand-site management	58.8%	23.5%	17.6%
Balancing energy cost estimation	70.6%	17.6%	11.8%
Production site analysis	64.7%	23.5%	11.8%

Among a variety of qualitative aspects which may be achieved by implementing reliable and accurate forecasting technology (compare Table 2), the majority of the participants considers an increased supply stability, a better possibility of balancing cost estimations and the identification of possible renewable energy production sites as the most important benefits. Surprisingly, only 35.3% see an importance of supply forecasts for smart grid applications while 41.2% oppose and 23.5% abstain. The most important user requirements for such solutions are the availability of appropriate statistical output evaluation measures, the robustness of the models in terms of their flexibility when adapting to changing situations and the necessary maintenance efforts (compare Table 3).

Table 3. Importance of quality aspects for forecasting software. Values are obtained by dichotomous distribution from 5-level scales.

Requirement	Important	Irrelevant	Abstention
Statistical error measures	92.9%	7.1%	0.0%
Technical performance	85.8%	7.1%	7.1%
Robustness/adaptability	92.9%	7.1%	0.0%
Application usability	71.4%	21.4%	7.1%
Maintenance efforts	92.9%	7.1%	0.0%
Graphical result representation	64.3%	35.7%	0.0%
Manual output pre-processing	71.4%	21.4%	7.1%

Regarding the forecast evaluation criteria preferences, the most important error measures for the users are the *Mean Absolute Percentage Error* (MAPE) with 24% and *Standard Deviation* (SD) with 20%, followed by MAE and RMSE with 16% each while 20% abstained. A comparison to the results obtained from literature review and software providers is shown in Table 4: MAPE and SD seem to be more suited for the industry, while researchers prefer to use RMSE and MAE. The absence of the MBE measure in the industry is surprising, because this means that differences between over- and underestimations are not considered as such important.

Table 4. Comparison of error measures used as forecast evaluation criteria in science and industry.

Scientific literature	Software providers	Software users
65% RMSE	25% SD	24% MAPE
47% MAE	21% MAPE	20% SD
28% MBE	21% RMSE	16% RMSE
25% MAPE	13% MAE	16% MAE
20% Others	13% Others	2% Others

4 Summary

Based upon the results presented in Sects. 2 and 3 we conclude that the scientific interest on renewable energy forecasting methods is unabated for more than one decade, making it a small but increasing research field among other popular forecasting topics related to the energy domain. Machine learning, hybrid- and multivariate time series regression models are the most frequently proposed solutions in the last 6 years covered by our study. Physical models are still under-represented but are advancing as part of hybrid models, while univariate methods do not have much significance. Commercial solutions mostly follow that direction with some exceptions, while their application areas are not explicitly restricted to the renewable energy domain. Although a method's quality is commonly considered as output accuracy and the RMSE is the dominating accuracy measure, the qualitative comparison of concurring methods remains difficult due to the lack of widely respected evaluation standards or reference models. Furthermore, the preferences for error measures differ between science and industry.

Acknowledgment. The work presented in this paper was funded by the European Regional Development Fund (EFRE) under co-financing by the Free State of Saxony and Robotron Datenbank-Software GmbH.

Appendix: List of Reviewed Journal Articles

	Journal Article	Year
SE	A 24-h forecast of solar irradiance using artificial neural network: Application for performance prediction of a grid connected PV plant at Trieste, Italy	2010
SE	A benchmark of statistical regression methods for short-term forecasting of photovoltaic electricity production. Part II: Probabilistic forecast of daily production	2014
SE	A hybrid model (SARIMA-SVM) for short-term power forecasting of a small-scale grid-connected photovoltaic plant	2013
RE	A hybrid model for wind speed prediction using empirical mode decomposition and artificial neural networks	2012
RE	A hybrid strategy of short term wind power prediction	2013
ECM	A new approach to very short term wind speed prediction using k-nearest neighbor classification	2013
ECM	A new wind speed forecasting strategy based on the chaotic time series modelling technique and the Apriori algorithm	2014
WEIA	A novel wind speed modeling approach using atmospheric pressure observations and hidden Markov models	2010
RSER	A review of measure-correlate-predict (MCP) methods used to estimate long-term wind characteristics at a target site	2013
RSER	A review of solar energy modeling techniques	2012
RE	A statistical approach for sub-hourly solar radiation reconstruction	2014
WEIA	A statistical method to merge wind cases for wind power assessment of wind farm	2013
TSTE	A Unified Approach for Power System Predictive Operations Using Viterbi Algorithm	2014
TSTE	A Weather-Based Hybrid Method for 1-Day Ahead Hourly Forecasting of PV Power Output	2014
ECM	Aggregated wind power generation probabilistic forecasting based on particle filter	2015
ECM	An adaptive model for predicting of global, direct and diffuse hourly solar irradiance	2010
RE	Bayesian adaptive combination of short-term wind speed forecasts from neural network models	2011
SE	Clear-sky irradiance predictions for solar resource mapping and large-scale applications: Improved validation methodology and detailed performance analysis of 18 broadband radiative models	2012
SE	Cloud motion and stability estimation for intra-hour solar forecasting	2015
ECM	Comparison of four Adaboost algorithm based artificial neural networks in wind speed predictions	2015
ECM	Comparison of new hybrid FEEMD-MLP, FEEMD-ANFIS, Wavelet Packet-MLP and Wavelet Packet-ANFIS for wind speed predictions	2015
SE	Daily global solar radiation prediction based on a hybrid Coral Reefs Optimization Extreme Learning Machine approach	2014
RE	Data mining and wind power prediction: A literature review	2012
TSTE	Determination Method of Insolation Prediction With Fuzzy and Applying Neural Network for Long-Term Ahead PV Power Output Correction	2013
SE	Embedded nowcasting method using cloud speed persistence for a photovoltaic power plant	2015
RE	Environmental data processing by clustering methods for energy forecast and planning	2011
SE	Estimation of solar radiation using a combination of Hidden Markov Model and generalized Fuzzy model	2013
WEIA	Estimation of wind speed: A data-driven approach	2010
SE	Evaluation and improvement of TAPM in estimating solar irradiance in Eastern Australia	2014
RSER	Evaluation of hybrid forecasting approaches for wind speed and power generation time series	2012
RE	Evolutive design of ARMA and ANN models for time series forecasting	2012
RSE	Feasibility study of a novel methodology for solar radiation prediction on an hourly time scale: A case study in Plymouth, United Kingdom	2014
SE	Forecasting solar radiation on an hourly time scale using a Coupled AutoRegressive and Dynamical System (CARDS) model	2013
SE	Genetic programming for photovoltaic plant output forecasting	2014
TSTE	Geostrophic Wind Dependent Probabilistic Irradiance Forecasts for Coastal California	2013
SE	Hourly solar irradiance time series forecasting using cloud cover index	2012
SE	Hybrid intra-hour DNI forecasts with sky image processing enhanced by stochastic learning	2013
SE	Hybrid solar forecasting method uses satellite imaging and ground telemetry as inputs to ANNs	2013
ER	Improved synthetic wind speed generation using modified Mycielski approach	2012
RE	Improved wind prediction based on the Lorenz system	2015
EPSE	Improving solar energy prediction in complex topography using artificial neural networks: Case study Peninsular Malaysia	2015
ECM	Intelligent optimization models based on hard-ridge penalty and RBF for forecasting global solar radiation	2015
RE	Local models-based regression trees for very short-term wind speed prediction	2015
TSTE	Model of Photovoltaic Power Plants for Performance Analysis and Production Forecast	2012
SE	Model output statistics cascade to improve day ahead solar irradiance forecast	2015
SE	Modeling global solar radiation using Particle Swarm Optimization (PSO)	2012
RE	Neural network approach to estimate 10-min solar global irradiation values on tilted planes	2013
ECM	Neural network based method for conversion of solar radiation data	2013
ECM	Observation and calculation of the solar radiation on the Tibetan Plateau	2012
RE	On the role of lagged exogenous variables and spatiotemporal correlations in improving the accuracy of solar forecasting methods	2015
SE	Online 24-h solar power forecasting based on weather type classification using artificial neural network	2011
RE	Online multi-step prediction for wind speeds and solar irradiation: Evaluation of prediction errors	2014
SE	Post-processing of solar irradiance forecasts from WRF model at Reunion Island	2014
ECM	Probabilistic wind power forecasting with online model selection and warped gaussian process	2014
SE	PV power forecast using a nonparametric PV model	2015
RE	Quaternion-valued short-term joint forecasting of three-dimensional wind and atmospheric parameters	2011
RE	Real-time prediction intervals for intra-hour DNI forecasts	2015
RSER	Selection of most relevant input parameters using WEKA for artificial neural network based solar radiation prediction models	2014
SE	Short-mid-term solar power prediction by using artificial neural networks	2012
RE	Short-term predictability of photovoltaic production over Italy	2015
RE	Short-term prediction of wind power with a clustering approach	2010
RE	Short-term solar power prediction using a support vector machine	2013
TSTE	Short-Term Wind Power Prediction Using a Wavelet Support Vector Machine	2012
WEIA	Simultaneous nested modeling from the synoptic scale to the LES scale for wind energy applications	2011
SE	Solar irradiance forecasting using a ground-based sky imager developed at UC San Diego	2014
SE	Solar irradiance forecasting using spatio-temporal empirical kriging and vector autoregressive models with parameter shrinkage	2014
TSTE	Solar Power Prediction Using Interval Type-2 TSK Modeling	2012

Journal Article		Year
RSER	Solar radiation forecasting with multiple parameters neural networks	2015
RSER	Solar radiation prediction using Artificial Neural Network techniques: A review	2014
RE	Solar radiation variability in Nigeria based on multiyear RegCM3 simulations	2015
SE	Stochastic approach for daily solar radiation modeling	2011
ECM	Stochastic models for wind speed forecasting	2011
IET	Study of forecasting renewable energies in smart grids using linear predictive filters and neural networks	2011
RE	The impact of large scale atmospheric circulation patterns on wind power generation and its potential predictability: A case study over the UK	2011
SE	The potential of different artificial neural network (ANN) techniques in daily global solar radiation modeling based on meteorological data	2010
TSTE	Time Adaptive Conditional Kernel Density Estimation for Wind Power Forecasting	2012
RSER	Validation of direct normal irradiance predictions under arid conditions: A review of radiative models and their turbidity-dependent performance	2015
SE	Very short term irradiance forecasting using the lasso	2015
RE	Very short-term wind speed forecasting with Bayesian structural break model	2013
RE	Wind power forecasting based on principle component phase space reconstruction	2015
WEIA	Wind power prediction based on numerical and statistical models	2013
RE	Wind speed forecasting in three different regions of Mexico, using a hybrid ARIMA-ANN model	2010
RE	Wind speed forecasting using a portfolio of forecasters	2014

References

1. Chen, Z., Yang, Y.: Assessing forecast accuracy measures. Technical report 2004–2010, Iowa State University, Department of Statistics & Statistical Laboratory (2004)
2. Colak, I., Sagiroglu, S., Yesilbudak, M.: Data mining and wind power prediction: a literature review. Renew. Energy **46**, 241–247 (2012)
3. Glassley, W., Kleissl, J., van Dam, C.P., Shiu, H., Huang, J., Braun, G., Holland, R.: Current state of the art in solar forecasting. Technical report, California Renewable Energy Collaborative (CREC) (2012)
4. Hong, T., Pinson, P., Fan, S.: Global energy forecasting competition 2012. Int. J. Forecast. **30**(2), 357–363 (2013)
5. Hyndman, R.J., Koehler, A.B.: Another look at measures of forecast accuracy. Int. J. Forecast. **22**(4), 679–688 (2006)
6. Kostylev, V., Pavlovski, A.: Solar power forecasting performance - towards industry standards. In: 1st International Workshop on the Integration of Solar Power into Power Systems, Aarhus, Denmark (2011)
7. Madsen, H., Kariniotakis, G., Nielsen, H., Nielsen, T., Pinson, P.: A protocol for standardizing the performance evaluation of short-term wind power prediction models. Anemos project, European Commission (2004)
8. Monteiro, C., Bessa, R., Miranda, V., Botterud, A., Wang, J., Conzelmann, G., et al.: Wind power forecasting: state-of-the-art 2009. Technical report, Argonne National Laboratory (ANL) (2009)
9. National Renewable Energy Laboratory. Transmission Grid Integration - Data and Resources. http://www.nrel.gov/electricity/transmission/data_resources.html. Accessed 16 July 2016
10. Tascikaraoglu, A., Uzunoglu, M.: A review of combined approaches for prediction of short-term wind speed and power. Renew. Sustain. Energy Rev. **34**, 243–254 (2014)
11. Ulbricht, R., Hahmann, M., Donker, H., Lehner, W.: Systematical evaluation of solar energy supply forecasts. In: Lee Woon, W., Aung, Z., Madnick, S. (eds.) Data Analytics for Renewable Energy Integration, pp. 108–121. Springer, Heidelberg (2014)
12. Yadav, A.K., Chandel, S.S.: Solar radiation prediction using artificial neural network techniques: a review. Renew. Sustain. Energy Rev. **33**, 772–781 (2014)

Measuring Stakeholders' Perceptions of Cybersecurity for Renewable Energy Systems

Stuart Madnick[1]([✉]), Mohammad S. Jalali[1], Michael Siegel[1],
Yang Lee[2], Diane Strong[3], Richard Wang[1], Wee Horng Ang[1],
Vicki Deng[1], and Dinsha Mistree[1]

[1] Massachusetts Institute of Technology, Cambridge, USA
smadnick@mit.edu
[2] Northeastern University, Boston, USA
[3] Worcester Polytechnic Institute, Worcester, USA

Abstract. Renewable energy systems need to be able to make frequent and rapid adjustments to address shifting solar and wind production. This requires increasingly sophisticated industrial control systems (ICS). But, that also increases the potential risks from cyber-attacks. Despite increasing attention to technical aspects (i.e., software and hardware) of cybersecurity, many professionals and scholars pay little or no attention to its organizational aspects, particularly to stakeholders' perceptions of the status of cybersecurity within organizations. Given that cybersecurity decisions and policies are mainly made based on stakeholders' perceived needs and security views, it is critical to measure such perceptions. In this paper, we introduce a methodology for analyzing differences in perceptions of cybersecurity among organizational stakeholders. To measure these perceptions, we first designed House of Security (HoS) as a framework that includes eight constructs of security: confidentiality, integrity, availability, technology resources, financial resources, business strategy, policy and procedures, and culture. We then developed a survey instrument to analyze stakeholders' perceptions based on these eight constructs. In a pilot study, we used the survey with people in various functional areas and levels of management in two energy and ICS organizations, and conducted a gap analysis to uncover differences in cybersecurity perceptions. This paper introduces the HoS and describes the survey instrument, as well as some of the preliminary findings.

1 Introduction

Rising demand for renewable energy resources has led to a noticeable focus on undertaking technological innovations to expand the green energy industry and respond to demand. As a result, cybersecurity has emerged as a critical issue as the green energy sector faces growing cyber risks. For example, smart grids—which are meant to provide reliable and efficient power network systems to distribute renewable energy resources—open up more direct and indirect connections to the Internet and more connections among the nodes in the networks. Smart grids also require advanced computing and communication technologies [1]. Adding new sources of renewable

© Springer International Publishing AG 2017
W.L. Woon et al. (Eds.): DARE 2016, LNAI 10097, pp. 67–77, 2017.
DOI: 10.1007/978-3-319-50947-1_7

energy to grids also requires an increase in the frequency and speed of technological adjustments. Consequently, while enhanced features and functionalities are introduced, the networked systems become increasingly vulnerable [2, 3]. Other complications and vulnerabilities are also added with the Internet of Things (IoT), where intelligent devices are getting connected, as sensors and/or controllers, within energy networks. In fact, not only are vulnerabilities on the rise, but they also have the potential of becoming very sophisticated, given the unknown characteristics of new technologies. Because a great deal of attention is being focused on technological innovations in renewable energy systems, the cybersecurity research community has also focused mostly on the technical aspects. Overall, a similar trend is observed in energy companies as they face the challenges of the high cost of developing new technologies in a context of limitation of available resources. As a result, it is not surprising that the organizational aspects of cybersecurity have become a blind spot for both industry and academia.

Cybersecurity is an increasingly crucial and complex management issue. Many organizations have developed cybersecurity policies to protect their business information and operational systems. Although these policies are important, they are often not fully adopted, the reasons being that organizations are limited by the resources they can devote to cybersecurity, and they often misunderstand the status of their cybersecurity. An organization's goal should be to develop the best possible, most cost-effective approach to cybersecurity, which is further complicated by the different priorities of organizational stakeholders. Stakeholders' perceptions of cybersecurity play a critical role in achieving this goal, since they are the main source of decision-making. Moreover, as organizations evolve into extended enterprises, which includes ties with suppliers, customers, and other partners, there is a significant increase in the number of stakeholders, and a wider range of security complications and requirements.

The current cybersecurity literature does not adequately address these issues. Many professionals and scholars have approached the study of cybersecurity by focusing specifically on the technical (e.g., hardware and software) and detailed elements of the security systems themselves, such as encryption [4–6], firewall technologies [7–9], and antiviruses [10, 11], or have measured specific events, such as mean-time-to-failure. Although these efforts are necessary, they often do not look at cybersecurity holistically and commonly neglect to consider its organizational aspects. They especially neglect to consider the perceived needs and security views of organizational stakeholders.

In this paper, we introduce the MIT House of Security (HoS) framework and present a survey instrument to measure stakeholders' perceptions of cybersecurity. We seek to identify similarities and differences, both within and between different organizations. This research has three major objectives:

- To identify how perceptions both shape, and should potentially shape, decisions about investments in security systems, with a particular focus on identifying the areas most in need of cybersecurity, as perceived by the individuals in the organization.
- To identify perceived differences between importance and assessment of the HoS constructs among stakeholders. These differences are further compared among individuals with different organizational roles and functional areas; e.g., comparing the views of mid-level managers to those of senior management, or the views of

information technology (IT) or operational technology (OT) workers with those of other members in the organization.

- To identify differences between the importance and assessment of the HoS constructs among different organizations (e.g., comparing two different organizations).

2 MIT's House of Security

Through a comprehensive literature review and several surveys, researchers at MIT have divided cybersecurity issues primarily into eight meta-groupings (i.e., constructs). Good security protects the "confidentiality" and "integrity" of data while providing "availability" of the data, networks, and systems to appropriate and authorized users. Confidentiality, integrity, and availability, also known as CIA, are often used as the only critical information characteristics [12]. Good security practices also go beyond just technical solutions and are driven by a "business strategy," with associated "policies and procedures" for security, and are implemented in a "culture of security." Moreover, these practices are supported by "technology resources" and "financial resources" dedicated to security. These eight constructs form the proposed House of Security and are shown in Fig. 1.

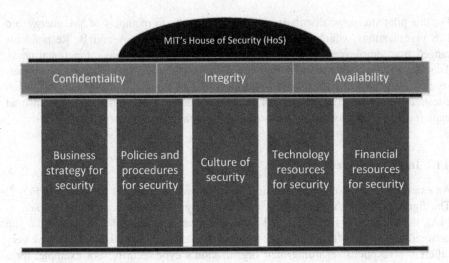

Fig. 1. The eight constructs of the House of Security

3 Survey Instrument

The survey includes three questions related to each construct of HoS (a total of 24 questions). In each question, respondents are asked to specify their perception of both the level of "assessment" and "importance." For example, they first respond to a question (e.g., "are people in the organization aware of good security practices?"), then identify the importance of that aspect. All questions are on a seven-point scale; "1" represents the smallest extent and "7" the largest extent.

The survey questions do not explicitly identify the construct being measured, but relate to aspects of the construct. Furthermore, there are multiple questions for each construct that are ordered randomly. The individuals are not aware of the categorization of the questions across the eight constructs.

A key part of this study involves gap analysis: how much does the perception of the current state of a cybersecurity aspect differ from the perception of its importance. Such gaps help identify potential opportunities for improvement within and across the extended enterprise. Differences in gaps among organizational stakeholders may represent different levels of understanding of security and help identify differences in local knowledge and needs.

We evaluated the quality of the survey instrument by measuring the statistical significance of the questions and the constructs and the reliability of the constructs by computing Cronbach's alpha [13]. While a key goal of our survey is to measure perceptions of the different constructs of security, we also plan to study the causes of these perception variations in our future research.

4   Preliminary Results of the Pilot Study

For this pilot study, we distributed the survey broadly to members of two energy and ICS organizations, which we will refer to as organizations A and B. Respondents ranged from employees to top-level managers and across all major functional areas. This diversity was important in order to identify variations in perceptions of cybersecurity. Here we briefly discuss some examples of the results based on: individual questions; constructs (i.e., a group of questions about an HoS construct); and construct gaps (i.e., the gap between assessment and importance of a construct).

4.1   Individual Questions

An example of the results of a question for organizations A and B are shown in Fig. 2. The figure presents the assessment of a user (my perceived assessment, marked as MA), the importance (my perceived level of importance, marked as MI), and the gap between MA and MI. This illustrates that people in different organization can have very different perceptions regarding their organization's cybersecurity. For example, for a question about well-defined and communicated cybersecurity strategy, there was a large gap (particularly in organization B), which implies that aspect falls far short of what is perceived to be needed among the respondents. Moreover, this example shows that organization A not only has a higher assessment about this question, but also they also have a higher expectation.

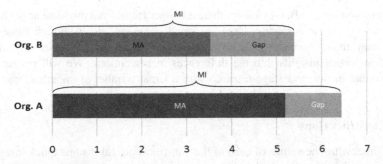

Fig. 2. Responses to a question on a seven-point scale: "The organization has a well-defined and communicated cybersecurity strategy." MA: my perceived assessment, MI: my perceived level of importance, Gap = MI-MA

4.2 Constructs

Beyond the individual questions, the results of the constructs present a more holistic overview. Each HoS construct contains three related questions, and the results of the questions are aggregated to present the construct. We have found, so far, that for a given organization, the assessment levels are likely to differ across the eight constructs, while the importance levels are often similarly high. Comparing organizations, one can observe and study both similarities and differences between the organizations.

The aggregated results of the eight constructs for organizations A and B are presented in Fig. 3[1]. The two organizations are relatively similar in their perceptions of availability, but differ noticeably in their perceptions of the state of policies and procedures for security—see Fig. 3. At this point, we are not focusing on the specifications

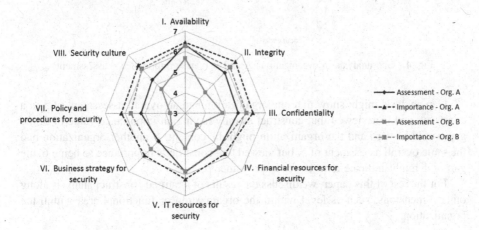

Fig. 3. Assessment vs. importance in organizations A and B

---

[1] Since assessment and importance values are usually above 4, we show the range 3 to 7 on the graph.

of organizations A and B. Obviously, there are other factors that might be at work, such as private vs. public company, large vs. small company, etc. Although these other factors may make it challenging to compare the organizations, these diagrams do provide important insights into the differences in perceptions. We will pursue these issues further in our next stage research with a larger number of organizations.

4.3   Construct Gaps

Although viewing the values of each of the constructs provides some quick insights, it is often more intuitive to examine the gaps between assessment and importance levels. The construct gaps in organizations A and B are presented in Fig. 4[2]. As can be seen, in this case, organization B has significantly larger gaps than organization A, with Policy and Procedures for Security construct having the largest gap.

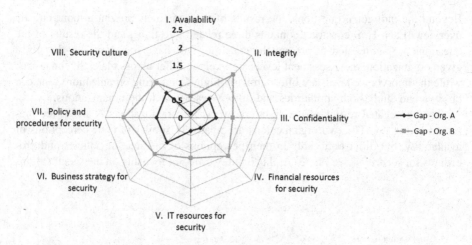

Fig. 4. Gap analysis in organizations A and B (gap = importance − assessment)

Gap analysis might show that one organization had an overall assessment of 5 in a construct, and if it viewed that construct as only having an importance value of 5, the gap would be zero and the organization might be content. If another organization had the same overall assessment of 5, but viewed that construct's importance as being 6, the gap of 1 might indicate an area for improvement.

For the rest of this paper, we discuss the results of stratified construct analysis along other dimensions, such as level within the organization or functional area within the organization.

---

[2] Since construct gaps are usually less than 2.0, we display gap values in multiples of 0.5 from 0 to 2.5.

5 Construct Analysis

Figure 5A shows the distribution of cybersecurity perceptions (i.e., construct assessment levels) based on the organizational level of the respondents: from executive level, to line managers, to professionals. Significant differences can be seen: Executives giving generally lower assessments, professions and middle managers in the middle, and "Others" with highest assessments.

Although ratings of assessment and importance are individually important, the size of the gaps can provide more insights (see Fig. 5B). The results show that top-level executives tend to have much higher gaps, across almost all constructs, than middle management and non-management personnel. This disparity in perceptions may imply that executives are more dissatisfied with the security situation in their organizations. Perhaps executives think situations are worse than they really are because they do not understand how and whether security measures are being correctly implemented. Or

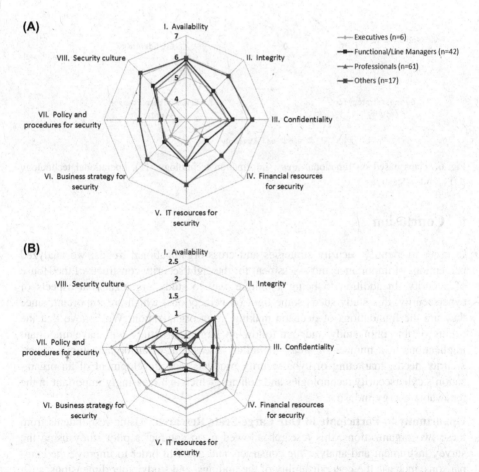

Fig. 5. Assessment levels (A) and gaps (B) by organizational levels in organizations.

alternatively, executives might see problems that people in other roles do not see and, as a result, their perceptions of a security gap are higher.

Overall, the sample sizes in this pilot study are small; hence, we use these findings to illustrate some of the issues that we expect will be significant in our larger study. We are currently conducting a large-scale study to better compare the results across various organizational levels. Follow-up studies and case studies would also help further clarify the underlying causes of differences in perceptions.

Figure 6 presents the gaps among IT, OT, and other areas in organization (such as Marketing, Finance, etc.). Interestingly, OT staff generally have higher gaps across the eight constructs. This is consistent with the frequent mention of IT/OT cultural gaps.

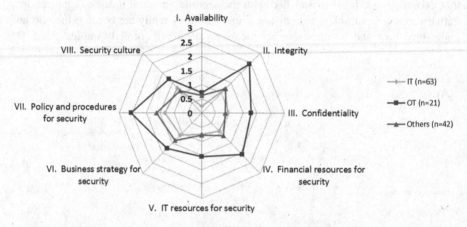

Fig. 6. Gaps based on functional areas: information technology (IT), operational technology (OT), and other areas

6 Conclusion

In order to identify security strategies and cross-organizational trends, we analyzed perceptions of importance and assessment for the eight security constructs of the House of Security. In addition to being a unique way to study organizational aspects of cybersecurity, this study sheds some light on perceptions, which are important, since they are the foundations of decision-making in an organization. We believe that the results of this pilot study and our follow-up large-scale study will have important implications in a number of areas, including assessment of an organization's cyber-security needs, marketing of cybersecurity products, and development of an organization's cybersecurity technologies and policies, which is increasingly important in the renewable energy industry.

Opportunity to Participate in Our Large-Scale Research: Using respondents from these two organizations, this research allowed us to conduct a pilot study using the survey instrument and analyze the constructs and gaps. In order to improve the comparisons, increase the generalizability of the findings, and study other dimensions, such as differences among industries, we are developing a larger dataset. We invite you and your organization to participate in our confidential organization benchmarking exercise, similar to organizations A and B in this paper. If you would like more information about this opportunity, please contact the corresponding author.

Acknowledgement and Disclaimer. This research was conducted by the MIT Interdisciplinary Consortium for Improving Critical Infrastructure Cybersecurity, also known as MIT-(IC)[3]. This material is based, in part, upon work supported by the Department of Energy under Award Number DE-OE0000780. We thank those who participated and provided the survey data. Early research was supported, in part, by Cisco Systems, Inc. through the MIT Center for Digital Business.

This report was prepared as an account of work sponsored, in part, by an agency of the US Government. Neither the US Government nor any agency thereof, nor any of their employees, makes any warranty, express or implied, or assumes any legal liability or responsibility for the accuracy, completeness, or usefulness of any information, apparatus, product, or process disclosed, or represents that its use would not infringe privately owned rights. The views and opinions of authors expressed herein do not necessarily state or reflect those of the US Government or any agency thereof.

Appendix

A. In my organization, I am a/an:

- Executive (CEO, CFO, VP, etc.)
- Functional or Line Manager
- Professional (Consultant, Engineer, In-house Expert, etc.)
- Other organizational member

B. In my organization, I work in area of:

- Information Technology (IT) Security
- IT, but not Security
- Operational Technology (OT) Security
- OT, but not Security
- General/Physical Security
- Business Security Policy or Management
- Other, i.e., not in Security, IT, or OT (e.g., Marketing, Accounting, HR, etc.), Please specify: _____

Assessment Scale:

1 = In my view, this security statement is true to a very SMALL extent in my organization.
7 = In my view, this security statement is true to a very LARGE extent in my organization.

Importance Scale:

1 = In my view, it is <u>NOT at all Important</u> that my organization address this security statement.

7 = In my view, it is <u>VERY Important</u> that my organization address this security statement.

1. The organization's business strategy sets direction for its cybersecurity practices.
2. The organization has adequate safeguards against internal and external threats to its data and networks.
3. In the organization, cybersecurity funds are appropriately distributed.
4. In the organization, the IT group takes cybersecurity seriously.
5. The organization's data and networks are available to approved users.
6. The organization has adequate policies for when and how data can be shared.
7. The organization has adequate technology for supporting cybersecurity.
8. People in the organization carefully follow good cybersecurity practices.
9. The organization has a well-defined and communicated cybersecurity strategy.
10. Cybersecurity is a funding priority in the organization.
11. The organization uses its IT security resources effectively to improve cybersecurity.
12. The organization has adequate policies about user identifications, passwords, and access privileges.

Assessment Scale:

1 = In <u>my view</u>, this security statement is true to a very <u>SMALL</u> extent in my organization.

7 = In <u>my view</u>, this security statement is true to a very <u>LARGE</u> extent in my organization.

Importance Scale:

1 = In my view, it is <u>NOT at all Important</u> that my organization address this security statement.

7 = In my view, it is <u>VERY Important</u> that my organization address this security statement.

13. The organization adequately monitors its data and networks against possible attacks.
14. Cybersecurity is a business agenda item for top executives in the organization.
15. The organization has well-defined policies and procedures for cybersecurity.
16. People in the organization can be trusted to engage in ethical practices with data and networks.
17. The organization has procedures for detecting and punishing cybersecurity violations.
18. In the organization, business managers help set the cybersecurity strategy.
19. The organization makes good use of available funds for cybersecurity.

20. The organization provides good access to data and networks to legitimate users.
21. The organization has a rapid response team ready for action when cyber attacks occur.
22. The organization protects its confidential corporate data.
23. People in the organization are aware of good cybersecurity practices.
24. The organization's data and networks are usually available when needed.
C. What is the biggest concern that you have about cybersecurity? (need not be included in the questions above) _____
D. Any other comments or suggestions? _____
E. If you would like to receive a copy of our research results, please provide your email address: (optional) Email: _____

We thank you for your time spent taking this survey.

References

1. Gharavi, H., Ghafurian, R.: Smart grid: the electric energy system of the future. IEEE (2011)
2. Liu, J., Xiao, Y., Li, S., Liang, W., Chen, C.P.: Cyber security and privacy issues in smart grids. IEEE Commun. Surv. Tutor. **14**, 981–997 (2012)
3. Pearson, I.L.: Smart grid cyber security for Europe. Energy Policy **39**, 5211–5218 (2011)
4. Boneh, D., Franklin, M.: Identity based encryption from the weil pairing. SIAM J. Comput. **32**, 586–615 (2003)
5. Dolev, D., Yao, A.: On the security of public key protocols. IEEE Trans. Inf. Theory **29**, 198–208 (1983)
6. Needham, R.M., Schroeder, M.D.: Using encryption for authentication in large networks for computers. Communications **21**, 993–999 (1978)
7. Cheswick, W.R., Bellovin, S.M., Rubin, A.D.: Firewalls and Internet Security: Repelling the Wily Hacker. Addison-Wesley, Boston (2003)
8. Oppliger, R.: Internet security: firewalls and beyond. Assoc. Comput. Mach. **40**, 92–103 (1977)
9. Zwicky, E., Cooper, S., Chapman, D., Ru, D.: Building Internet Firewalls. O'Reilly & Associates, Sebastopol (2000)
10. Furnell, S.: Cyber threats: what are the issues and who sets the agenda? In: SGIR Conference (2004)
11. Kephart, J., Sorkin, G., Chess, D., White, S.: Fighting Computer Viruses. Sci. Am. **277**, 88–93 (1997)
12. McCumber, J.: Assessing and Managing Security Risk in IT Systems. Auerbach Publications, Boca Raton (2005)
13. Cronbach, L.J.: Coefficient alpha and the internal structure of tests. Psychometrika **16**, 297–334 (1951)

Selection of Numerical Weather Forecast Features for PV Power Predictions with Random Forests

Björn Wolff[1(✉)], Oliver Kramer[2], and Detlev Heinemann[1]

[1] Institute of Physics, Energy Meteorology,
Carl von Ossietzky University in Oldenburg, Oldenburg, Germany
bjoern.wolff@uni-oldenburg.de
[2] Department of Computing Science, Computational Intelligence,
Carl von Ossietzky University in Oldenburg, Oldenburg, Germany

Abstract. The increasing volatility introduced to power grids by renewable energy sources makes it necessary that the accuracy of energy forecasts are improved. Photovoltaic (PV) power plants hold the biggest share of installed capacity of renewable energy in Germany, so that high quality PV power forecasts are vital for a cost efficient operation of the underlying electrical grid. In this paper, we evaluate multiple Numerical Weather Prediction (NWP) parameters for their ability to improve PV power forecasting features. The importance of features is decided by a Random Forest algorithm. Furthermore, the resulting top ranked features are tested by performing PV power forecasts with Support Vector Regression, Random Forest, and linear regression models.

Keywords: PV power forecasting · Random Forest · Feature importance · Support Vector Regression

1 Introduction

One major part of ensuring the stability of a electricity grid is keeping a fixed utility frequency, which rises with production and lowers with the consumption of energy. Until about a decade ago, transmission system operators solved this problem by simply matching the energy output and dispatch of power plants to the demand of its consumers. Today, the increasing share of weather dependent renewable energy sources on the consumer's side has been introducing new volatility to the grid, thus, keeping stability intact is becoming a more complex task. In Germany, photovoltaic (PV) systems take up the highest share of installed capacity of renewable energy. PV is able to reach power output rates of almost 26 GW at midday with a total nominal power of around 40 GW_{peak} [25]. To generate the same amount of energy one would need about 20 of the bigger nuclear power plants in Germany, whereas today only eight plants with varying capacities are still active. To integrate this amount of energy, and still ensure

© Springer International Publishing AG 2017
W.L. Woon et al. (Eds.): DARE 2016, LNAI 10097, pp. 78–91, 2017.
DOI: 10.1007/978-3-319-50947-1_8

grid stability, in a cost efficient way, i.e., without relying on too much expensive reserve energy, accurate power forecasts for PV are necessary.

One widely used approach for PV power forecasting is based on parametric modeling of PV power using measurements and numerical weather predictions (NWP) on varying spatial and temporal scales. A comparison of models was compiled by Pelland et al. [22]. In recent years, more and more statistical learning algorithms were used for wind and solar energy predictions. One popular approach for forecasting solar radiation are artificial neural networks (ANN). A compilation of different ANN models can be found in Mellit [18]. More recent works that focus on ANN are extending the models with preprocessing steps such as weather type classifications [5], feature selection implemented with, e.g., Genetic Algorithms [21] or increasing the input data by adding meteorological data from numerical weather predictions [17]. There are also many works successfully applying and comparing different statistical modeling approaches for PV energy production forecasting [2,16,19].

In comparison to ANN, Support Vector Regression (SVR) is still less common for PV power and irradiance forecasts but shows good potential in some comparisons to other statistical learning methods [27] with numerical weather prediction features as input [13] on single-site and regional solar radiation values [7]. There are only a few works that use Random Forest (RF) as a predictor in the field of renewable energy forecasts. One example where RF is successfully used for electric load forecasting can be found in Jurado et al. [11].

Here, we evaluate Random Forest's ability to assess the importance of numerical weather prediction features for PV power forecasting. The datasets used in this evaluation and the preprocessing of our data is described in Sect. 2. In Sect. 3, we give a short overview on the statistical learning methods utilized in this work for PV power predictions, i.e., Random Forest and Support Vector Regression. The selection of numerical weather forecast features is conducted in Sect. 4 and the most important features are evaluated with different regression models in Sect. 5.

2 Datasets

In this work, we use the data of 93 PV systems spread across Germany (see Fig. 1) with varying installed capacities from rooftop installations to big PV parks. These systems are monitored by our industry partner meteocontrol GmbH[1] and are randomly selected from a larger dataset containing PV power measurements from 2012-01-01 to 2013-12-31. The temporal resolution of these measurements is 15 min and the specifications, i.e., orientation, tilt, and power capacity are known.

2.1 Numerical Weather Predictions

The focus of this work lies on evaluating the usefulness of different weather parameters of a numerical weather prediction (NWP) system for PV power

---

[1] meteocontrol GmbH: www.meteocontrol.com.

forecasting. NWP models are systems of differential equations that use the laws of physic, fluid dynamics, and chemistry as well as weather observations and measurements as a basis to predict changes in weather situations starting from a recent initial state. These calculations are performed on a fixed horizontal and vertical grid.

Here, we use data from the European Centre for Medium-Range Weather Forecasts (ECMWF)[2]. The ECMWF's Integrated Forecasting System (IFS) provides single level forecasts covering over 120 parameters in a temporal resolution of three hours with two major forecasting runs starting at 00 UTC and 12 UTC. Single level forecasts are forecasts that are averaged over different height levels or forecasts for specific heights (e.g., 2 metre temperature). These runs are not instantly available at these times, so that we always use forecasts from the 12 UTC run of the previous day to ensure their availability at the point of time of our prediction. In Fig. 1, the grid points of the ECMWF IFS model are shown with each grid field covering an area of about 12.5 km × 12.5 km. As we only use single level forecasts, there is no need for the vertical expansion of the grid in our case.

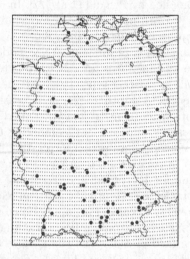

Fig. 1. Locations of 93 PV systems in Germany (red) and grid points of ECMWF's NWP model (blue) (Color figure online)

2.2   Data Preprocessing

The NWP model output is interpolated both temporally and spatially to match the PV systems' measurements.

Spatial Interpolation. In previous works (see Lorenz et al. [14]), a positive effect of averaging NWP radiation forecasts of multiple surrounding grid points instead of using the geographically nearest forecast was observed. The best results were achieved by averaging a 4 × 4 grid around a PV system's location. Due to the easier implementation and faster calculation, the spatial interpolation in this work is done with a distance weighted k-nearest neighbor regression model [1] with $k = 16$ in regard to the 4 × 4 grid. For consistency, this method of spatial interpolation is further applied for all other NWP forecasts used in this work.

Temporal Interpolation. In case of irradiance forecast parameters, we are utilizing a clear sky model of Dumortier, described in Fontoynont et al. [6], to interpolate the data from three hours to 15 min. For all other parameters, we

---

[2] European Centre for Medium-Range Weather Forecasts: www.ecmwf.int.

apply a normal linear interpolation. The clear sky interpolation is working in three steps:

1. Calculating clear-sky index k_{3h}^* for 3 h values:

$$k_{3h}^* = \frac{I_{forec,3h}}{I_{clearsky,3h}}$$

2. Linear interpolation of k_{3h}^* to 15 min values results in $k_{15\,min}^*$.
3. Using $k_{15\,min}^*$ as a factor for $I_{clearsky,15\,min}$ with 15 min resolution:

$$I_{forec,15\,min} = k_{15\,min}^* \cdot I_{clearsky,15\,min}.$$

With these interpolations, we are now able to generate PV power forecasts with a resolution of 15 min for each of PV system using the following statistical learning approaches.

3 Statistical Learning Methods

In this section, we introduce the applied statistical learning methods, i.e., Random Forest for feature selection and Support Vector Regression for modeling PV power forecasts. While the two methods construct a regression function f mapping patterns X_i, consisting of one or more features, to labels y_i in different ways, both algorithms need a dataset Z containing known pattern and label combinations for training purposes. The mapping can be defined as follows:

$$\begin{pmatrix} X_1 \\ X_2 \\ \dots \\ X_N \end{pmatrix} = \begin{pmatrix} x_{11} & x_{12} & \dots & x_{1d} \\ x_{21} & x_{22} & \dots & x_{2d} \\ \dots & \dots & \dots & \dots \\ x_{N1} & x_{N2} & \dots & x_{Nd} \end{pmatrix} \xrightarrow{f} \begin{pmatrix} y_1 \\ y_2 \\ \dots \\ y_N \end{pmatrix},$$

with pattern $X_i \in \mathbb{R}^d$, where d denotes the number of features, label $y_i \in \mathbb{R}$, and training set size $N \in \mathbb{N}$.

3.1 Random Forest

The concept behind the construction of a Random Forest (RF) [3] regressor is Bagging (Bootstrap aggregation). Bagging makes it possible to "average many noisy but approximately unbiased models, and hence reduce the variance" [10] by combining multiple single models. With RF this variance reduction through correlation reduction of the trees is further improved by selecting input features in each node (one element of the tree) splitting step at random.

To train a Random Forest model, bootstrapping (random sampling with replacement) is applied on the training set Z to retrieve B, the number of trees in the *forest*, subsets $Z^{*b} = \{(\mathbf{x}_1^*, y_1^*), (\mathbf{x}_2^*, y_2^*), \dots, (\mathbf{x}_N^*, y_N^*)\}$, $b = 1, 2, \dots, B$. On each of these subsets a RF tree T_b is grown by applying the following steps until a stop criterion is reached (e.g., a minimum number of samples belonging to a newly created node), according to Hastie et al. [10]:

1. Randomly select $m \leq d$ features of the input pattern.
2. Calculate the best feature for splitting, i.e., the feature that "maximizes the decrease of some impurity measure" [15].
3. Split the node according to the selected feature into two new nodes.

After finishing the tree-growing process, the algorithm outputs an ensemble of trees $\{T_b \mid b = 1, 2, \ldots, B\}$ and the Random Forest regression function is

$$f_{rf}^B(\mathbf{x}) = \frac{1}{B} \sum_{b=1}^{B} T_b(x). \tag{1}$$

In our implementation, we use the Random Forest regressor method of the *scikit-learn* [23] Python package with its standard settings except for the number of trees. We increased the number of trees from ten to 64 to improve tree diversity but still keep the computation time at a minimum. This is as well the minimal recommendation of Oshiro et al. [20], even though they have a different field of use. As a impurity measure (step 2 of the tree-growing algorithm) to decide on the best splitting criterion/feature, this implementation uses the mean square error (mse)

$$E_{mse} = \frac{1}{N} \sum_{i=1}^{N} (z - z')^2. \tag{2}$$

The E_{mse} for each possible feature split is calculated and the feature with the highest E_{mse} decrease is selected.

3.2   Support Vector Regression

The basic idea of Support Vector Regression [24] is to find the regression function f that maps patterns to labels by solving the optimization problem

$$\inf_{f \in \mathcal{H}, b \in \mathbb{R}} \frac{1}{N} \sum_{i=1}^{N} \mathcal{L}_\epsilon \big(y_i, f(\mathbf{x}_i + b)\big) + \lambda \|f\|_{\mathcal{H}}^2. \tag{3}$$

Here, $\lambda \in \mathbb{R} > 0$ is a fixed user-defined cost parameter and the ϵ-insensitive loss function $\mathcal{L}_\epsilon$ is defined as $\mathcal{L}_\epsilon(y, t) = \max(0, |y - t| - \epsilon), \epsilon \in \mathbb{R} > 0$. $\|f\|_{\mathcal{H}}^2$ describes the squared norm in a so-called reproducing kernel Hilbert space $\mathcal{H}$ induced by an associated kernel function k: $\mathcal{X} \times \mathcal{X} \to \mathbb{R}$. The space $\mathcal{H}$ contains all considered models, and the term $\|f\|_{\mathcal{H}}^2$ is a measure for the complexity of a particular regression model f [10]. Because of good results in related publications, e.g., [7,12,27], we use the radial basis function (RBF) kernel requiring another parameter $\gamma \in \mathbb{R} > 0$.

In summary, there are three user-defined parameters, i.e., λ, ϵ, and γ that we optimize by applying grid search with 80 different parameter combinations. Again, we use the SVR implementation of *scikit-learn* which is based on *LIB-SVM* [4].

4   Feature Selection

The main goal of this work is to find additional NWP weather parameters that improve PV power forecasting. Using all available weather parameters of the ECMWF's model would result in feature spaces with over 120 dimensions. This is not feasible for short-term PV power forecasting with forecast horizons of 15 min as the time to calculate a forecast would simply take too much time. As a result, we want to select less features that still hold the highest possible additional information for our models.

In literature (e.g., in Guyon and Elisseeff [9]), there are two major classes of feature selection methods: wrapper and filter. While wrapper methods embed a regression model into another optimization algorithm, e.g., genetic algorithms, that try different feature combinations until a stop-criterion is reached and select the best solution that was found along the search, filter methods rank single features by some measure that describe their usefulness for the given task.

As one iteration of a wrapper algorithm would take a lot of time in our scenario with 93 PV systems, we decided to use a filter model to achieve a fast method for evaluating the benefit of single weather parameters. Instead of using common measures like Pearson's correlation coefficient, we apply a comparatively new approach that is the build-in feature importance ranking of the Random Forest algorithm.

Preselection of Features. With the help of experts in the field of meteorology, we removed all NWP parameters that are irrelevant for PV power forecasting. Thus, from the about 120 available IFS single level forecast parameters, 30 features remain for our evaluation. These 30 features are listed in Table 1.

Feature Selection with Random Forest. To rank the preselected features, we use Random Forest's feature importance capabilities. The feature importance is achieved by traversing all trees with the training dataset and summing the impurity criterion (in this case the E_{mse} value) of each node. The E_{mse} is weighted with the number of samples that were routed to a specific node. Thus, features that were choosen often and early in a trees hierarchy will also receive high importance values. The output of this feature importance algorithm is normalized to one. Details on Random Forest's feature importance are well discussed in Genuer et al. [8].

Now, all preselected features are presented to the RF algorithm with the goal of calculating a regression function to predict power output of each single PV system. For this, we use the daytime (cosine of the solar zenith angle above 80 degree) feature values of the previous 65 days for training as a similar configuration showed good results in previous evaluations [26]. Each of the 93 PV systems is trained independently for each day of the year 2012 starting at 2012-03-10. By doing this, we receive the feature importance of every day separately and are able to evaluate seasonal or daily changes in feature importances.

In Fig. 2, the feature importance of all days is averaged over all PV systems for summer and winter months. In summer, all features are dominated by the

Table 1. Selection of ECMWF weather forecast feature list. Marked features (*) are interpolated with a clear sky interpolation method instead of a linear interpolation.

No.	Feature name	Unit
1	100 metre U wind component	ms^{-1}
2	100 metre V wind component	ms^{-1}
3	10 metre U wind component	ms^{-1}
4	10 metre V wind component	ms^{-1}
5	2 metre dewpoint temperature	K
6	2 metre temperature	K
7	Clear-sky direct solar radiation at surface*	Jm^{-2}
8	Cloud base height	m
9	Evaporation	m of water equivalent
10	High cloud cover	$(0-1)$
11	Large-scale precipitation	m
12	Low cloud cover	$(0-1)$
13	Medium cloud cover	$(0-1)$
14	Snow density	kgm^{-3}
15	Snow depth	m of water equivalent
16	Snow evaporation	m of water equivalent
17	Snowfall	m
18	Surface net solar radiation, clear sky*	Jm^{-2}
19	Surface solar radiation*	Jm^{-2}
20	Surface solar radiation downwards*	Jm^{-2}
21	Total cloud cover	$(0-1)$
22	Total column ice water	kgm^{-2}
23	Total column liquid water	kgm^{-2}
24	Total column rain water	kgm^{-2}
25	Total column snow water	kgm^{-2}
26	Total column water	kgm^{-2}
27	Total column water vapour	kgm^{-2}
28	Total precipitation	m
29	Total sky direct solar radiation at surface*	Jm^{-2}
30	Zero degree level	m

forecast of surface solar radiation (19) and surface solar radiation downwards (20) with over 50% feature importance. Only the parameters total sky direct solar radiation at surface (29) and evaporation (9) seem to hold some information for the resulting PV power output with an importance of about 7.5% each. This changes when looking at the winter months. The feature importances of 19 and

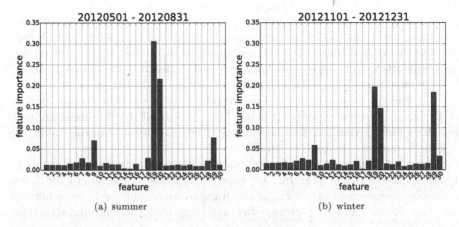

Fig. 2. Random Forest feature importances of 30 NWP weather parameters averaged for (a) summer and (b) winter months for all 93 PV systems.

Table 2. Top ten most important ECMWF weather forecast features ranked by Random Forest feature importance.

Rank	No.	Feature name	Feature importance
1	19	Surface solar radiation	0.244
2	20	Surface solar radiation downwards	0.178
3	29	Total sky direct solar radiation at surface	0.171
4	9	Evaporation	0.050
5	7	Clear-sky direct solar radiation at surface	0.026
6	18	Surface net solar radiation, clear sky	0.023
7	8	Cloud base height	0.020
8	30	Zero degree level	0.019
9	6	2 metre temperature	0.018
10	16	Snow evaporation	0.017

20 decrease by about one third while the importance of many other features increase. Noticeably, the zero degree level (30) and low cloud cover (12) as well as snow depth (14) and snow evaporation (15) become more important for the forecasting of PV power generation in winter.

In our following experiments, we want to evaluate whether adding weather parameters to the prediction pattern can improve the quality of PV power forecasts. In this first approach, we do not differentiate between seasons and use the ten highest ranked features of the average feature importance for 2012. These features are listed ranked by their RF feature importance values in Table 2.

5   Prediction Comparison

Now, we use the most important features obtained in Sect. 4 for PV power forecasts. In this evaluation, we train our models at each day of the year 2013 and our training set consists of the previous 65 days as before. The training and forecasting is done on each PV system separately. Aside from the Random Forest predictor used to determine the feature importance, we test a linear regression and Support Vector Regression approach and evaluate if the Random Forest feature importance is applicable for different learning algorithms as well.

To measure the quality of the different predictors, we use the E_{rmse}, the root of the E_{mse} introduced in Eq. 2. The E_{rmse} is a good real life measure as it penalizes high deviations, that would have a higher impact on grid stabilizing actions in our case, more than small ones.

Instead of testing all possible combinations, we decided to iteratively add the top ten features to our pattern. Our base pattern (Eq. 4) consists of the latest available measurement values P_{meas} according to the considered forecast horizon Δt. The different applied pattern/label matchings have the following structure:

$$(P_{meas}(t - \Delta t)) \rightarrow P_{meas}(t) \quad (4)$$
$$(P_{meas}(t - \Delta t), Feature_1(t)) \rightarrow P_{meas}(t) \quad (5)$$
$$\ldots$$
$$(P_{meas}(t - \Delta t), Feature_1(t), Feature_2(t), \ldots, Feature_{10}(t)) \rightarrow P_{meas}(t) \quad (6)$$

Figure 3 shows the results of our forecasts for the linear regression, SVR, and RF models. First, we look at the shortest forecast horizon of 15 min in Fig. 3(a). The E_{rmse} values of the forecast models using only measurement data (features 0) are already good and can not profit much from additional features. Especially after adding the NWP forecast of surface solar radiation, which was deemed the most important feature in our importance tests, there is almost no further improvement of the E_{rmse} for the linear and SVR model anymore. Only Random Forest forecasts acquire their best results with the top five features but the E_{rmse} values are still higher than that of the other two models.

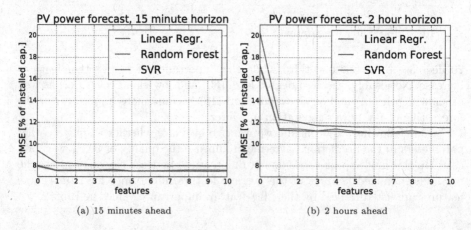

(a) 15 minutes ahead   (b) 2 hours ahead

Fig. 3. Comparison of Linear Regression, Random Forest, and Support Vector Regression with increasing input feature count for a prediction horizon of (a) 15 min and (b) two hours ahead.

When the temporal difference between the most recent measurement and the forecasted timestep gets bigger, the less important the measurements get. This is demonstrated in Fig. 3(b) for a two hour forecast horizon. The initial measurement-based forecasts achieve a higher E_{rmse} for a forecast horizon of two hours than in the 15 min case. As before, adding only the most important feature is already enough to reach the lowest E_{rmse} value for the linear regression and SVR (only very slight improvements afterwards). Again, the Random Forest model needs more information, i.e., more features to reach its best forecast with the lowest E_{rmse}.

In both scenarios, the difference between a simple linear regression and the more sophisticated Support Vector Regression with optimized user-defined parameters is small. In Fig. 4, we look at the most important feature, i.e., surface solar radiation, which allows the highest E_{rmse} improvement, in more detail. Here, the average of all stations' measurements and the respective average of the surface solar radiation forecasts is compared. The distribution shows that these parameters hold a strong linear dependency, which is highlighted by a linear regression fit in this figure. This indicates that the SVR is not able to benefit from its ability to model non-linear relations and, therefore, is uncapable of achieving lower E_{rmse} values than a linear regression.

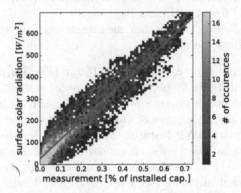

Fig. 4. Scatter plot of measurements and corresponding NWP forecast of surface solar radiation. The red line indicates the linear regression curve of these two variables. (Color figure online)

Due to the fact that there was almost no improvement in adding more than one NWP feature, we try different input combinations in Fig. 5. Because there was not much of a difference between the results of the linear and the SVR model as well as the shorter computation times of a linear regression, we decided on testing these combinations only with the linear model. We test three different models:

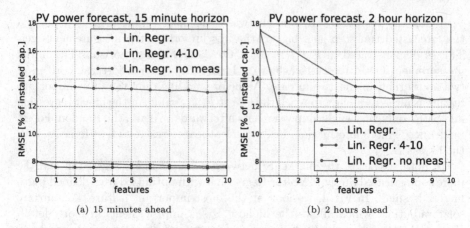

(a) 15 minutes ahead   (b) 2 hours ahead

Fig. 5. Comparison of Linear Regression forecasts with different input feature combinations for a prediction horizon of (a) 15 min and (b) two hours ahead

- *Model 0*: Linear Regression with measurements and all top ranked features as before (Linear Regr.)
- *Model 1*: Linear Regression with measurements and without the top three ranked (radiation) features (Linear Regr. 4–10)
- *Model 2*: Linear Regression without measurements and with all top ranked features (Linear Regr. no meas)

In case of our shortest forecast horizon of 15 min (Fig. 5(a)), the models using measurements are substantially better than the model without measurement values. While there are no improvements of the E_{rmse} after adding feature 1 for Model 0, the three missing features of Model 1 can be (almost completely) recuperated by adding more NWP features. Although the forecasts of Model 2 are inferior to measurement-based ones, an improvement is achieved by adding more of the top ten features. For a forecast horizon of two hours ahead, measurements are not that important anymore (see Fig. 5(b)). Now, Model 2, as the forecast does not change with time due to the same NWP forecast used, is competitive to Model 0 and even better than Model 1, despite the lack of measurements. There are slight differences of Model 2's E_{rmse} values for 15 min and two hours as we filter the time series in a way that only time steps where all models generate useable outputs are considered, so that the E_{rmse} values are calculated on different slightly timeseries. Model 1 is profiting from additional NWP features and able to compensate less information about the predicted radiation with increasing feature count. Utilizing all information (Model 0) is still generating the best forecasts, but with an increasing forecast horizon the difference to the other models is increasingly vanishing.

6   Conclusion

The expansion of PV power in the German grid makes it necessary that its forecasts become more accurate. To address this task, we evaluate the benefit of using additional weather forecast parameters of the ECMWF's NWP model for PV power predictions applying different regression models. The importance of all features that are related to the output of a PV system are assessed via Random Forest's feature importance algorithm. While there are differences between seasons and weather situations, the averaged results of the feature importance evaluation show that over 50% of importance are shared by irradiance weather parameters. This is later on confirmed by our analysis of PV power forecasts, that are using the highest ranked features, showing that the quality of power output forecasts for more than a few minutes ahead is mostly depending on irradiance forecasts. Otherwise, for shorter prediction horizons (15 to 30 min), PV power measurements are essential for generating high quality forecasts. These results are seen with all of the applied regression models, i.e., Support Vector Regression (SVR), Random Forest, and linear regression. As irradiance and a PV module's power output are highly linear correlated, SVR is not able to create better forecasts than the linear regression approach.

In further works, we want to investigate whether there are weather situations where additional features actually increase the forecast quality, e.g., snow and fog.

Acknowledgements. We like to thank meteocontrol GmbH for providing the PV system measurements as well as the ECMWF for numerical weather prediction forecasts that are basis of our experimental analysis. Björn Wolff is funded by the PhD program "System Integration of Renewable Energies" (SEE) of the University of Oldenburg promoted by the Lower Saxony Ministry for Science and Culture (MWK).

References

1. Bailey, T., Jain, A.K.: A note on distance-weighted k-nearest neighbor rules. IEEE Trans. Syst. Man Cybern. **8**, 311–313 (1978)
2. Brabec, M., Pelikán, E., Krc, P., Eben, K., Musilek, P.: Statistical modeling of energy production by photovoltaic farms. In: 2010 IEEE Electric Power and Energy Conference (EPEC), pp. 1–6, August 2010
3. Breiman, L.: Random forests. Mach. Learn. **45**(1), 5–32 (2001)
4. Chang, C.-C., Lin, C.-J.: LIBSVM – a library for support vector machines (2015). Website, last checked 11 Feb 2016
5. Chen, C., Duan, S., Cai, T., Liu, B.: Online 24-h solar power forecasting based on weather type classification using artificial neural network. Sol. Energy **85**(11), 2856–2870 (2011)
6. Fontoynont, M., Dumortier, D., Heinnemann, D., Hammer, A., Olseth, J., Skarveit, A., Ineichen, P., Reise, C., Page, J., Roche, L., et al.: Satellight: a WWW server which provides high quality daylight and solar radiation data for Western and Central Europe. In: 9th Conference on Satellite Meteorology and Oceanography, pp. 434–437. American Meteorological Society, Boston (1998)

7. Gala, Y., Fernández, Á., Díaz, J., Dorronsoro, J.R.: Support vector forecasting of solar radiation values. In: Pan, J.-S., Polycarpou, M.M., Woźniak, M., Carvalho, A.C.P.L.F., Quintián, H., Corchado, E. (eds.) HAIS 2013. LNCS (LNAI), vol. 8073, pp. 51–60. Springer, Heidelberg (2013). doi:10.1007/978-3-642-40846-5_6
8. Genuer, R., Poggi, J.-M., Tuleau-Malot, C.: Variable selection using random forests. Pattern Recogn. Lett. 31(14), 2225–2236 (2010)
9. Guyon, I., Elisseeff, A.: An introduction to variable and feature selection. J. Mach. Learn. Res. 3, 1157–1182 (2003)
10. Hastie, T., Tibshirani, R., Friedman, J.: The Elements of Statistical Learning. Springer, Berlin (2001)
11. Jurado, S., Peralta, J., Nebot, A., Mugica, F., Cortez, P.: Short-term electric load forecasting using computational intelligence methods. In: 2013 IEEE International Conference on Fuzzy Systems (FUZZ), pp. 1–8, July 2013
12. Kramer, O., Gieseke, F.: Short-term wind energy forecasting using support vector regression. In: Corchado, E., Snásel, V., Sedano, J., Hassanien, A.E., Calvo, J.L., Ślęzak, D. (eds.) SOCO 2011. Advances in Intelligent and Soft Computing, vol. 87, pp. 271–280. Springer, Heidelberg (2011)
13. Krömer, P., Musílek, P., Pelikán, E., Krc, P., Jurus, P., Eben, K.: Support vector regression of multiple predictive models of downward short-wave radiation. In: 2014 International Joint Conference on Neural Networks, IJCNN 2014, Beijing, China, 6–11 July 2014, pp. 651–657 (2014)
14. Lorenz, E., Hurka, J., Heinemann, D., Beyer, H.G.: Irradiance forecasting for the power prediction of grid-connected photovoltaic systems. IEEE J. Select. Top. Appl. Earth Observ. Remote Sens. 2(1), 2–10 (2009)
15. Louppe, G., Wehenkel, L., Sutera, A., Geurts, P.: Understanding variable importances in forests of randomized trees. In: Burges, C.J.C., Bottou, L., Welling, M., Ghahramani, Z., Weinberger, K.Q. (eds.) Advances in Neural Information Processing Systems, vol. 26, pp. 431–439. Curran Associates Inc., Red Hook (2013)
16. Mandal, P., Madhira, S.T.S., Haque, A.U., Meng, J., Pineda, R.L.: Forecasting power output of solar photovoltaic system using wavelet transform and artificial intelligence techniques. Procedia Comput. Sci. 12, 332–337 (2012)
17. Marquez, R., Coimbra, C.F.: Forecasting of global and direct solar irradiance using stochastic learning methods, ground experiments and the NWS database. Sol. Energy 85(5), 746–756 (2011)
18. Mellit, A.: Artificial intelligence technique for modelling and forecasting of solar radiation data: a review. Int. J. Artif. Intell. Soft Comput. 1, 52 (2008)
19. Monteiro, C., Santos, T., Fernandez-Jimenez, L., Ramirez-Rosado, I., Terreros-Olarte, M.: Short-term power forecasting model for photovoltaic plants based on historical similarity. Energies 6(5), 2624–2643 (2013)
20. Oshiro, T.M., Perez, P.S., Baranauskas, J.A.: How many trees in a random forest? In: Perner, P. (ed.) MLDM 2012. LNCS (LNAI), vol. 7376, pp. 154–168. Springer, Heidelberg (2012). doi:10.1007/978-3-642-31537-4_13
21. Pedro, H.T., Coimbra, C.F.: Assessment of forecasting techniques for solar power production with no exogenous inputs. Sol. Energy 86(7), 2017–2028 (2012)
22. Pelland, S., Remund, J., Kleissl, J., Oozeki, T., De Brabandere, K.: Photovoltaic and solar forecasting: state of the art (2013). http://www.iea-pvps.org/index.php?id=1&eID=dam_frontend_push&docID=1690
23. scikit learn. Machine learning in Python (2016). Website, last checked 11 Feb 2016
24. Smola, A.J., Schölkopf, B.: A tutorial on support vector regression. Stat. Comput. 14, 199–222 (2004)

25. Wirth, H., Schneider, K.: Recent facts about photovoltaics in Germany, April 2016.   http://www.ise.fraunhofer.de/en/renewable-energy-data/data-and-facts-about-pv
26. Wolff, B., Kühnert, J., Lorenz, E., Kramer, O., Heinemann, D.: Comparing support vector regression for PV power forecasting to a physical modeling approach using measurement, numerical weather prediction, and cloud motion data. Sol. Energy **135**, 197–208 (2016)
27. Zeng, J., Qiao, W.: Short-term solar power prediction using a support vector machine. Renew. Energy **52**, 118–127 (2013)

Evolutionary Multi-objective Ensembles
for Wind Power Prediction

Justin Heinermann[1(✉)], Jörg Lässig[2], and Oliver Kramer[1]

[1] Department of Computing Science, University of Oldenburg, Oldenburg, Germany
{justin.heinermann,oliver.kramer}@uni-oldenburg.de
[2] Department of Computer Science,
University of Applied Sciences Zittau/Görlitz, Zittau, Germany
jlaessig@hszg.de

Abstract. Ensembles turn out to be excellent wind power prediction methods. But the space of algorithms and parameters of supervised learning ensembles is large. For an efficient optimization and tuning of ensembles, we propose to employ evolutionary multi-objective optimization methods in this work. NSGA-II is a classic optimization algorithm based on non-dominated sorting and maximization of the crowding distance and has successfully been applied in various applications in the past. The experimental part of the paper shows how NSGA-II tunes SVR ensembles, random forests, and heterogenous ensembles. The study demonstrates that the proposed approach evolves an attractive set of ensembles for a practitioner yielding numerous compromises of prediction accuracy and runtime.

1   Introduction

Statistical learning methods allow precise wind power predictions for short forecast horizons [7]. The success depends on the choice of algorithms and parameter settings, which can be a quite tedious task. Ensembles of regression techniques help to decrease the prediction error, often with an acceptable computational time. However, the problem of a growing number of parameters and algorithmic choices is even more difficult for ensembles. In this paper, we propose to overcome the curse-of-dimensionality problem for parameters with evolutionary multi-objective optimization algorithms, which find a tradeoff between runtime and prediction error.

Balancing machine learning methods with evolutionary multi-objective optimization is an active research field, e.g., with the non-dominated sorting genetic algorithm NSGA-II by Deb *et al.* [2]. The evolution of a set of alternative machine learning models is an appealing approach for the practitioner, who can choose his favorite model a posteriori. Further, the optimization with multi-objective approaches is cheaper. It requires much less evaluations for a certain number of balanced models than the same number of separate weighted single-objective runs would require. Numerous previous research results demonstrate the success of balancing machine learning models. Mierswa [9] balanced regularization and

© Springer International Publishing AG 2017
W.L. Woon et al. (Eds.): DARE 2016, LNAI 10097, pp. 92–101, 2017.
DOI: 10.1007/978-3-319-50947-1_9

accuracy of support vector machines with NSGA-II. Hu et al. [6] propose an evolutionary method for decision tree ensembles. In [10] we balanced accuracy and runtime with NSGA-II on artificial benchmark datasets.

In this paper we apply evolutionary balancing to ensembles for the prediction of wind power. Basis is a data-based regression approach using measurements of wind power in the environment of a target turbine. It employs regression techniques for mapping from the current wind power situation to the future. The prediction time horizon depends on the temporal and spatial resolution of the wind power data. The wind speed must be related to the spatial dimensions of the surrounding turbines as well as to the temporal resolution of the time series data. In [13] we use nearest neighbor regression as machine learning model, while we employ neural networks in [12].

This paper is structured as follows. In Sect. 2, we introduce the evolutionary multi-objective optimization algorithm NSGA-II, which is basis of the ensemble optimization approach. The paper concentrates on support vector regression (SVR) ensembles in Sect. 3, on random forests in Sect. 4, and on heterogenous ensembles in Sect. 5. Section 6 summarizes the most important results and gives insights into prospective future work.

2 NSGA-II

Evolutionary computing is a paradigm in computer science, which is inspired by natural evolution. There are different variants of evolutionary algorithms. The basic idea is to invoke populations of individuals in an iterative manner, see Eiben and Smith [3]. Each individual consists of a chromosome, whose phenotype achieves a fitness value w.r.t. an objective function. The chromosome $\mathbf{x} \in \mathbb{R}^N$ is a point in the space of possible solutions and can be seen as genotype of the individual. For a minimization problem, we seek for the solution with the lowest fitness value in the solution space. The candidate solutions in a population compete with each other and are evolved using crossover, mutation, and selection.

Evolutionary algorithms can be employed to solve multi-objective problems [3]. In contrast to a single-objective problem, the quality of a solution is defined by multiple objectives $f_1, \ldots, f_m$. Often, objectives are conflictive and a compromise is required. Optimal solutions in such a multi-objective optimization scenario are not worse than other solutions in all objectives. This concept is known as Pareto optimality and is based on the concept of dominance. Solution $\mathbf{x}$ dominates solution $\mathbf{x}'$:

$$\mathbf{x} \prec \mathbf{x}', \tag{1}$$

if it is better in all objectives $f_1, \ldots, f_m$. We seek for a set of non-dominated solutions that defines a Pareto front in objective space

$$\mathcal{PF} = \{\mathbf{f}(\mathbf{x}^*) \in \mathbb{R}^m \mid \nexists \mathbf{x} \in \mathbb{R}^N : \mathbf{x} \prec \mathbf{x}^*\}. \tag{2}$$

Various algorithms have been proposed in the past for solving multi-objective problems. One of the most famous ones is NSGA-II by Deb *et al.* [2], which we use in this work. It is based on non-dominated sorting that computes a domination rank for each solution. For the actual population, the non-dominated solutions are identified getting rank 0. After these solutions are removed from the population, the new non-dominated solutions in the current set of solutions are determined getting the next higher rank. The process is repeated until the set of solutions is empty.

In the second step NSGA-II minimizes the crowding distance metric to achieve a smooth approximation of the Pareto front among the solutions with the lowest rank. The crowding distance is the sum of differences of single fitness values of the neighboring solutions. A large value means there are only few other solutions in the neighborhood of the solution. We use a $(\mu + \mu)$ population scheme, i.e., in each generation μ offspring solutions are generated. The parental and the offspring populations are merged and the new population is evolved based on dominance rank and crowding distance using a tournament operator [3]. Through elitism and diversity maintenance, the NSGA-II strategy performs comparatively successful.

3   SVR Ensembles

Support vector techniques belong to the most successful methods in machine learning and have also proven well in wind power prediction in the past. In [4], we showed the superiority of ensembles using SVR as base predictors over single SVR models. First, the prediction error can be decreased. Second, the runtime can be decreased utilizing the runtime behavior of the SVR algorithm and a divide & conquer approach. The best performing models were created using a random parameter choice for regularization parameter C and kernel bandwidth σ of an RBF-kernel for each SVR model. The reason for this is the diversity amongst the predictors.

In the following, we give a proof-of-concept that multi-objective optimization is a feasible approach for the model selection problem. This allows for balancing the runtime and the prediction error. We choose a simple way for constructing the ensemble resulting in a chromosome with six genes. The bagging ensemble consists of $T_1+T_2+T_3$ SVR models with different parameter setting and uniform weighting:

- T_1 models use samples of size S_1 and parameters $C = 1.0$ and $\sigma = 1.0$
- T_2 models using samples of size S_2 and parameters $C = 100.0$ and $\sigma = 10^{-3}$
- T_3 models using samples of size S_3 and parameters $C = 10000.0$ and $\sigma = 10^{-5}$

The basic idea is to use models of different complexity, which can be provided by different C and σ choices and sample sizes S_i. The number of possibilities for different SVR ensemble parameterizations are countless and therefore we only demonstrate the feasibility of the approach with this simple subset. Each T_i of the initial population is initialized with a random number from $\{1, \ldots, 50\}$, S_i from $\{1, \ldots, 2000\}$.

The experimental setup is the same for all studies throughout this paper. We employ three National Renewable Energy Laboratory (NREL) [8] test turbines from the Mojave desert using data from the year 2004, i.e., turbines from Lancaster (ID: 2473), Palm Springs, (ID: 1175), and Las Vegas (ID: 6272). The data consists of power measurements for each target turbine and its neighbors in 10-min steps. The dataset is split up with 5-fold cross-validation from `scikit-learn` [11].

The optimization process is conducted with NSGA-II using SBX-recombination and polynomial mutation [2]. In each experiment, we use population size $\mu = 20$. The termination condition is the maximum of 500 fitness function evaluations resulting in 25 generations. The objectives are (1) the training time and (2) the mean squared error (MSE) employing cross-validation (CV). The results of the optimization are visualized as scatter plot with training time on the x-axis, CV error on the y-axis and the generation number as color.

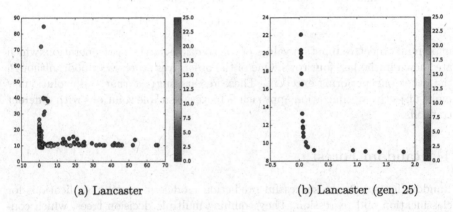

(a) Lancaster   (b) Lancaster (gen. 25)

Fig. 1. SVR ensemble solutions evolved by NSGA-II in 25 generations. (Color figure online)

The results for a wind turbine in Lancaster is shown in Fig. 1. While Fig. 1(a) shows all evaluated solutions, Fig. 1(b) shows the evolved approximation of the Pareto front in the 25th generation. We observe that a lot of different solutions with different error behavior and runtimes are generated. In each iteration the solutions are moving to the left bottom corner and therefore towards a desirable solution with a tradeoff between runtime and error. The solutions of the final generation, see Fig. 1(b), show a useful Pareto front for the machine learning practitioner, from which he can select a solution based on his preferences.

In Table 1 a selection of interesting solutions is presented. For training time t and CV error E, the minimum, median, and maximum in the 25th generation are shown and the corresponding solutions are depicted. While an optimal training time or an optimal prediction accuracy can only be achieved by neglecting the other objective, all tradeoffs achieve worse fitness values, but are a possibly preferable compromise. The same observation can be made for the minimum and

Table 1. Selected SVR ensemble solutions for Lancaster.

Pareto set	T_1	S_1	T_2	S_2	T_3	S_3	t	E
min(t)	0	0	0	191	17	41	0.03	22.06
median(t)	0	0	12	213	18	55	0.14	9.87
max(t)	1	0	18	733	26	444	1.83	9.10
min(E)	0	0	18	789	26	399	1.48	9.03
median(E)	0	0	12	213	18	55	0.14	9.87
max(E)	0	0	0	191	17	41	0.03	22.06
All solutions	T_1	S_1	T_2	S_2	T_3	S_3	t	E
min(t)	0	0	0	191	17	41	0.03	22.06
max(t)	44	1967	43	1653	48	1741	65.80	10.35
min(E)	0	0	18	767	29	438	1.80	9.01
max(E)	31	1	0	792	18	368	2.55	84.63

maximum objective function values of the solutions in the last generation, which are shown in the last four rows. One of the objectives scores very poor, although the other one performs excellent. These results suggest that the evolutionary multi-objective optimization approach achieves desirable solutions with different tradeoffs.

4   Random Forest

Random forests [1] are successful prediction models in various applications for classification and regression. They combine multiple decision trees, which conduct step-wise decisions for labels in a graph structure. Besides the number of estimators there are some decision tree-specific parameters like the tree depth or the maximum number of features considered for random splits. These have impact on both the prediction accuracy and the training time.

The results for random forests are very similar to the SVR ensembles. We tune the number of estimators that are initialized with a random number from $\{1, \ldots, 500\}$, the maximum features used for the random split, initially from $\{1, \ldots, 10\}$, and the maximum depth of the trees, initially from $\{1, \ldots, 5\}$. The results for the test wind turbine in Lancaster are shown in Fig. 2. By tuning the parameters and in particular adding more estimators to the ensembles, the prediction performance is improved.

In Table 2 we show some important solutions from the Pareto set and from all evaluated individuals. It is possible to select a solution that performs fast but yields a large error. In contrast to the min(t) solution, for a relatively good prediction error it is not necessary to spend the largest training time. The median(t) and median(E) solutions show that it is sufficient to invest only little more time to achieve an acceptable prediction error.

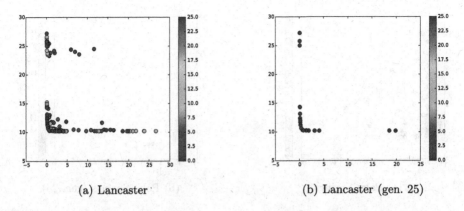

(a) Lancaster

(b) Lancaster (gen. 25)

Fig. 2. Random forest solutions evolved by NSGA-II in 25 generations.

Table 2. Selected random forest solutions for Lancaster.

Pareto set	n_estimators	max_features	max_depth	t	E
min(t)	10	1	1	0.07	26.31
median(t)	10	4	4	0.39	10.59
max(t)	296	7	4	18.24	10.21
min(E)	296	7	4	18.24	10.21
median(E)	10	4	4	0.40	10.65
max(E)	10	1	1	0.07	26.31
All solutions	n_estimators	max_features	max_depth	t	E
min(t)	10	1	1	0.06	27.21
max(t)	415	9	4	33.65	10.24
min(E)	315	7	4	19.86	10.19
max(E)	10	1	1	0.07	27.23

5   Heterogeneous Ensembles

Heterogeneous ensembles turned out to be exceedingly successful for the wind power prediction problem [5]. The heterogenous ensembles we employ in our study comprise SVR, k-nearest neighbors (kNN), and decision trees (DT). As ensembles benefit from the diversity of predictors, we showed that ensembles employing different prediction algorithms perform well and offer a moderate requirement of computational time for both training and testing.

In the following experiment we give an example for the optimization using NSGA-II for heterogeneous ensemble predictors for three wind turbines. The number of possible models is very large and we again choose a simple subset as proof-of-concept. In the experiment the predictors consist of $T_{SVR}+T_{DT}+T_{kNN}$ models. The parameters T_i are tuned with NSGA-II along with the corresponding sample sizes S_{SVR}, S_{DT}, S_{kNN}. Each T_i of the initial population is initialized

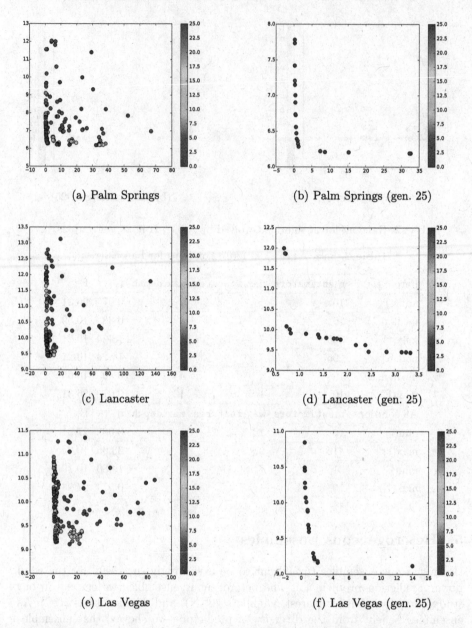

Fig. 3. Heterogeneous ensemble solutions evolved by NSGA-II in 25 generations.

with a random number from $\{1, \ldots, 100\}$, S_i from $\{1, \ldots, 2000\}$. The SVR models use an RBF-Kernel with $\sigma = 10^{-4}$ and $C = 1000$. k-NN uses $k = 5$ nearest neighbors.

Figure 3 shows the results for three test turbines near Palm Springs, Lancaster, and Las Vegas. For each turbine, all the 500 solutions depicted in the

Table 3. Selected heterogeneous ensemble solutions for Lancaster.

Pareto set	T_{SVR}	S_{SVR}	T_{DT}	S_{DT}	T_{kNN}	S_{kNN}	t	E
min(t)	3	121	0	223	100	1753	0.79	10.19
median(t)	3	156	3	332	25	1836	1.81	11.95
max(t)	48	124	67	242	108	1728	3.84	9.48
min(E)	55	147	6	224	111	1884	3.79	9.47
median(E)	3	125	0	226	108	80	1.16	10.11
max(E)	2	117	3	224	25	70	0.93	12.01
All solutions	T_{SVR}	S_{SVR}	T_{DT}	S_{DT}	T_{kNN}	S_{kNN}	t	E
min(t)	3	117	2	224	25	60	0.67	12.00
max(t)	82	1379	79	1972	20	398	133.55	11.80
min(E)	56	124	116	226	118	1755	3.32	9.44
max(E)	10	1698	22	839	7	1523	19.44	13.13

left plot show various possible solutions. The right plots shows the Pareto front evolved in the 25th generation. For the turbine in Lancaster, we depict the parameters and objective function values for interesting solutions in Table 3. Also for heterogenous ensembles, the balancing of runtime and prediction error can be solved with NSGA-II. It turns out that a quite heterogeneous composition of the ensemble is beneficial.

In Fig. 4 we present a comparison of the evolved SVR ensembles, random forests, and heterogeneous ensembles for the test turbine in Lancaster. Although it might not be a completely fair comparison because of some simplified assumptions, the results show that all approaches achieve very good solutions. The optimal SVR ensemble solutions achieve the best prediction error in the shortest

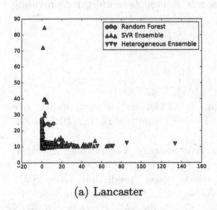

(a) Lancaster

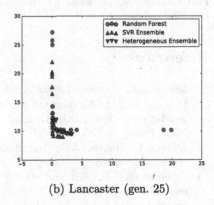

(b) Lancaster (gen. 25)

Fig. 4. Random forest (red), SVR ensemble (blue), and heterogeneous ensemble (green) solutions evolved by NSGA-II in 25 generations. (Color figure online)

time. However, both solution sets of the random forests and the SVR ensembles contain solutions with relatively poor prediction error. In contrast, the Pareto front evolved in the heterogeneous ensemble experiment only contains solutions that provide a good balance between the two objectives.

6   Conclusions

Evolutionary multi-objective algorithms like the famous NSGA-II heuristic that we used in our experimental study turn out to be successful balancing approaches for ensembles in wind power prediction applications. When dealing with supervised learning, not only the prediction performance is objective to parameter tuning, but also the reduction of long training times to a feasible level. Our results have shown that the approach successfully evolves a Pareto front of preferable solutions, from which the practitioner can select a balance of prediction accuracy and runtime. Very successful balancing results were achieved with the heterogenous ensemble variants.

The presented experiments serve as a proof-of-concept. Future work may concentrate on extensions of free parameters and an analysis of further ensemble methods, neural network ensembles. Also other parameters can be considered like ensemble weights, a selection of neighboring turbines, and further algorithmic parameters. This surely increases the search space of parameters, but also increases the flexibility of runtime and accuracy. Further, the combination of the data-driven approaches with meteorological predictions will probably lead to powerful ensembles. The integration of such predictors into an ensembles requires further parameter tuning without an influence on the runtime of the physical simulation models.

Acknowledgement. We thank the ministry of science and culture of Lower Saxony for supporting us with the PhD Program *System Integration of Renewable Energies*. Furthermore, we thank the US *National Renewable Energy Laboratory* for providing the wind dataset.

References

1. Breiman, L.: Random forests. Mach. Learn. **45**(1), 5–32 (2001)
2. Deb, K., Pratap, A., Agarwal, S., Meyarivan, T.: A fast and elitist multiobjective genetic algorithm: NSGA-II. IEEE Trans. Evol. Comput. **6**(2), 182–197 (2002). IEEE Press
3. Eiben, A., Smith, J.E.: Introduction to Evolutionary Computing. Springer, Heidelberg (2015)
4. Heinermann, J., Kramer, O.: Precise wind power prediction with SVM ensemble regression. In: Wermter, S., Weber, C., Duch, W., Honkela, T., Koprinkova-Hristova, P., Magg, S., Palm, G., Villa, A.E.P. (eds.) ICANN 2014. LNCS, vol. 8681, pp. 797–804. Springer, Heidelberg (2014). doi:10.1007/978-3-319-11179-7_100

5. Heinermann, J., Kramer, O.: Machine learning ensembles for wind power prediction. Renew. Energy **89**, 671–679 (2016). Elsevier
6. Hu, Q.-H., Yu, D.-R., Wang, M.-Y.: Constructing rough decision forests. In: Ślęzak, D., Yao, J.T., Peters, J.F., Ziarko, W., Hu, X. (eds.) RSFDGrC 2005. LNCS (LNAI), vol. 3642, pp. 147–156. Springer, Heidelberg (2005). doi:10.1007/11548706_16
7. Kramer, O., Gieseke, F., Satzger, B.: Wind energy prediction and monitoring with neural computation. Neurocomputing **109**, 84–93 (2013)
8. Lew, D., Milligan, M., Jordan, G., Freeman, L., Miller, N., Clark, K., Piwko, R.: How do wind and solar power affect grid operations: the western wind and solar integration study. In: 8th International Workshop on Large Scale Integration of Wind Power and on Transmission Networks for Offshore Wind Farms, pp. 14–15 (2009)
9. Mierswa, I.: Controlling overfitting with multi-objective support vector machines. In: Genetic and Evolutionary Computation Conference (GECCO), pp. 1830–1837 (2007)
10. Oehmcke, S., Heinermann, J., Kramer, O.: Analysis of diversity methods for evolutionary multi-objective ensemble classifiers. In: Mora, A.M., Squillero, G. (eds.) EvoApplications 2015. LNCS, vol. 9028, pp. 567–578. Springer, Heidelberg (2015). doi:10.1007/978-3-319-16549-3_46
11. Pedregosa, F., Varoquaux, G., Gramfort, A., Michel, V., Thirion, B., Grisel, O., Blondel, M., Prettenhofer, P., Weiss, R., Dubourg, V., Vanderplas, J., Passos, A., Cournapeau, D., Brucher, M., Perrot, M., Duchesnay, E.: Scikit-learn machine learning in Python. J. Mach. Learn. Res. **12**, 2825–2830 (2011)
12. Stubbemann, J., Treiber, N.A., Kramer, O.: Resilient propagation for multivariate wind power prediction. In: International Conference on Pattern Recognition Applications and Methods (ICPRAM), pp. 333–337 (2015)
13. Treiber, N.A., Kramer, O.: Evolutionary feature weighting for wind power prediction with nearest neighbor regression. In: IEEE Congress on Evolutionary Computation (CEC), pp. 332–337. IEEE Press (2015)

A Semi-automatic Approach for Tech Mining and Interactive Taxonomy Visualization

Ioannis Karakatsanis[✉], Alexandros Tsoupos, and Wei Lee Woon

Electrical Engineering and Computer Science,
Masdar Institute of Science and Technology, Abu Dhabi, United Arab Emirates
ikarakatsanis@masdar.ac.ae

Abstract. For research directors and other stakeholders, being able to identify emerging technologies and evaluate the comparative advantages and future potentials of these technologies is the highest importance. In previous work, we had proposed a fully automatic, taxonomy-based framework for identifying technologies that are in the early stages of growth and for visualizing their interrelationships. Although this method was very promising, it was apparent that when using a fully automatic process that some level of subjectivity and inconsistency was difficult to avoid. The current work addresses these shortcomings by developing a semi-automatic platform that would allow the incorporation of expert feedback into the tech mining process. To achieve this, a unified web-based application was implemented which combines the analytical techniques proposed in the previous studies with an interactive visualization experience. The proposed approach is evaluated by domain experts and appears to be capable of generating informative and accurate visualizations of early growth technologies.

Keywords: Technology forecasting · Semi-automatic · Interactive visualization · Renewable energy

1 Introduction

The ability to monitor and anticipate emerging trends is of paramount importance to science and technology decision makers [19]. Bibliometric methods, which are based on the statistics of publication metadata, are particularly valuable for technology forecasting (TF) in research fields consisting of many subfields and underlying technologies [21]. For this reason, we previously developed a unified bibliometric framework for identifying technologies that are in the early stages of growth and for visualizing the "research landscapes" in which these technologies are embedded.

1.1 Motivation and Objectives

The approach proposed in [24] was designed to be fully automatic and could already produce some very interesting results. However, in practice the resulting taxonomies, while generally accurate, did occasionally connect technologies

© Springer International Publishing AG 2017
W.L. Woon et al. (Eds.): DARE 2016, LNAI 10097, pp. 102–115, 2017.
DOI: 10.1007/978-3-319-50947-1_10

which were only distantly related [23]. Figure 1 shows an example of the kinds of problematic subtrees which were encountered. In this case, the term "chandipura virus" was incorrectly assigned to this subtree even though there was no obvious relevance.

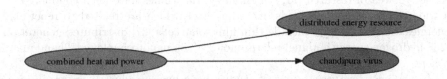

Fig. 1. Static visualization of a problematic subtree for technology "combined heat and power" (Color figure online)

Generated taxonomies can be interpreted in many different ways; in many cases it can be argued that there is no single "best" taxonomy and that the validity of a taxonomy or of the relationships therein depends greatly on the requirements and perspective of the user. Also, knowledge discovery is a naturally interactive and iterative process whereas most data visualizations are inherently static like the visualization produced by the aforementioned framework. Examination of the framework requires direct modification of the algorithmic parameters which can be somewhat indirect and very challenging for non-technical users. In addition, the large size of the taxonomies means that the effect of modifying these parameters can often be very difficult to determine. Since data science and its tools are becoming more and more widespread, there is need for tools that allow users to directly interact with the data [13].

Hence, the aim of this study is to overcome the aforementioned limitations by introducing a semi-automatic approach for taxonomy generation that yields more accurate results for tech-mining. The objectives of the proposed framework can be summarized as follows:

(a) To allow the users to incorporate prior knowledge into the taxonomy generation process by adjusting some algorithmic parameters.
(b) To introduce dynamic and interactive visualizations of keywords taxonomies that allow users to explore the data for themselves. This feature will be useful for new and expert users alike.
(c) To provide a web-based platform that will accessible by anyone using through a modern web browser, regardless of the operating system and device type.

To demonstrate the validity and benefits of the proposed methodology an in-depth evaluation of our approach is undertaken by domain experts where semi-automated taxonomies are compared with those produced using the fully automatic process.

2   Background

2.1   Related Work

The detection of emerging technologies using bibliometric methods and tools is an active area of research [11,15]. Bibliometric studies have been conducted on a wide range of research fields, including ones which are related to renewable energies. Some examples include thin film solar cells [27] distributed generation [26], hydrogen energy and fuel cell technology [28], nanotechnology [12] and many others.

In the past few years, a lot of studies have been conducted on graph visualizations [5,6,8]. From the perspective of interactive visualization, some of the frameworks developed in [3,7,16] are similar to the application presented here. However, both web-based frameworks in [3,7] focus on the fast representation of large scale graphs comprising of thousands of nodes. On the other hand, the application developed in [16] allows rich customization by changing the size and color of nodes, the strength of edges and adding textual descriptions to entities.

Most existing graph visualization tools are desktop based (for e.g. Cytoscape [17] and Pajek [1]). Nevertheless, the recent popularity of the World Wide Web (WWW) as a platform for the deployment of complex applications has resulted in a broad shift towards rich internet applications (RIAs) capable of offering interactive and responsive interfaces through a web browser [7]. Popular examples of such applications include Tom Sawyer Visualization [18] and Touchgraph [20].

There are currently very few studies for automatically visualizing and predicting the future evolution of areas of research. Thus, in [24] a fully automatic taxonomy-based framework for TF and visualization was proposed. It is important to note that the form of "forecasting" that this study was centered around was not in the sense of a weather forecast, where specific future predictions had to be made with a quite high level of certainty. Instead, the proposed framework was focused on finding promising but less obvious technological trends and bringing these to the attention of the human experts. Although this framework produced very interesting taxonomies, these were sometimes difficult to interpret [23]. Taxonomies are traditionally composed of "is-a" relationships, whereas the meanings of the links generated using the automatic procedure were a little harder to determine (for e.g. Fig. 1). There are various reasons for this - overly general terms which can be placed in multiple points in a taxonomy, incomplete and noisy observations such as those obtained from publicly available and locally cached databases, ambiguities in the meanings of some terms, incapability of the fully automatic process to identify semantic relationships rather than actual technological dependencies, etc. As a result, inconsistencies between the keyword nodes of the taxonomy occur.

2.2   Automatic Taxonomy Generation

The taxonomy generation framework on which the proposed methodology is based will now be introduced.

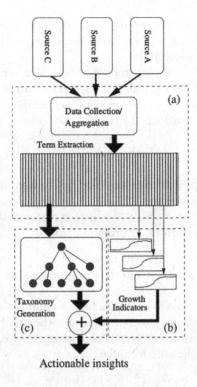

Fig. 2. Analytical framework [24]

Figure 2 depicts the high-level structure of the system proposed in [24]. The goal of this system is to be a comprehensive technology analysis mechanism which can collect data, extract relevant terms and statistics, estimate growth indicators and finally integrating these with keyword taxonomies to produce actionable results. The system has been divided into three segments:

1. Data collection and term extraction (labeled (**a**) in the figure).
2. Prevalence estimation and calculation of growth indicators (labeled (**b**)).
3. Taxonomy generation and integration with growth indicators (labeled (**c**)).

For an extensive description of the framework and its components, the interested reader is referred to [24].

Taxonomy Generation. An important capability of the framework presented in [24] is the automated creation of keyword taxonomies based on the idea that terms which co-occur frequently should be grouped together. To achieve this the algorithm described in [9] was used. To adapt the algorithm to the task of creating technological taxonomies, the asymmetric distance function proposed in [25] is used. This in turn is based on the "Google distance" proposed in [4]:

$$\overrightarrow{NGD}(t_x, t_y) = \frac{log\, n_y - log\, n_{x,y}}{log\, N - log\, n_x} \qquad (1)$$

where t_x and t_y are the two keywords being considered and n_x, n_y and $n_{x,y}$ are the occurrence counts for the two keywords occurring individually, then together in the same document respectively.

The algorithm consists of two steps: the first is to build a similarity matrix of keywords from which the centrality of each node can be calculated. Next, the taxonomy is grown by inserting the keywords sequentially in order of decreasing centrality. Each node t_i is attached to one of the nodes in the taxonomy t_j, such that:

$$j = \underset{j \in T}{\operatorname{argmin}} \overrightarrow{NGD}(t_i, t_j) \tag{2}$$

where T is the set of indices corresponding to terms which have already been incorporated into the taxonomy.

In addition, to maintain stability two further customizable optimizations were added to the basic algorithm as described in [23]:

1. Attachment of a node to a parent node is based on a weighted average of the similarities to the parent but also to the grandparents and higher ancestors of that node.
2. In some cases it was necessary to include a child penalty whereby the cost of attaching to a given parent increases once the number of children of that parent exceeds a certain number. The cost c_{t_j} of a candidate parent node increases using the following formula:

$$c_{t_j} = c_{t_j} \times \frac{1 + cp}{1 + e^{(p - n_{t_j})}} \tag{3}$$

where cp is the child penalty parameter which takes values in the range of $[0, 1]$, p is the maximum number of child nodes allowed before the penalty starts and n_{t_j} is the number of the children of node t_j. Initially, the cost c_{t_j} of a node is estimated as described earlier using the weighted average of the similarities to the parent but also to the grandparents and higher ancestors of that node.

Growth Indicators and Visualization. To derive an indicator of the growth potential for a technology, two elements are required. The first is a measure for the "prevalence" of a given area as function of time. An appropriate metric which seems to work acceptably is the term frequency, defined as:

$$TF_i = \frac{n_i}{\sum_{j \in x} n_j} \tag{4}$$

where n_i is the number of occurrences of keywords $i \in I$ and I is the set of terms occurring in all article abstracts.

The growth indicators are then calculated as follows:

$$\theta_i = \frac{\sum_{t \in y_1, y_2} t \times TF_i[t]}{\sum_{t \in y_1, y_2} TF_i[t]} \tag{5}$$

where θ_i is the growth potential for keyword i and $TF_i[t]$ is the term frequency for term i and year t while y_1 and y_2 are the first and last years in the study period. As it can be seen, the growth indicators actually represent the average publication year for articles appearing over the range of years being used and which are relevant to term i. In particular, a more recent year implies greater currency of the topic.

Once the keyword taxonomies have been created a simple method for enriching the early growth indicators using information regarding the co-occurrence statistics of keywords is used. The main idea is to re-calculate the early growth scores for each keyword based on the aggregate scores of each of the keywords occurred in the subtree descended from the corresponding node in the taxonomy. For the results obtained using the method presented in this paper, aggregation was achieved by simply averaging the respective nodes' scores together with the scores of all child nodes.

Finally, for representing the growth potentials a straightforward color-coding scheme is adopted. "Hot" technologies are coded red and there is a range of colors leading up to "cold" technologies in blue. Figure 1 shows an example of this scheme which is also caused by the method presented in the following section.

3   Proposed Methodology

In this section, a description of the user interface of the proposed web-based framework is provided along with technical implementation details.

3.1   User Interface

Initially, the user interface of our tool provides a simple grid of form controls in a horizontal layout where the user can adjust the parameters for the taxonomy generation. Specifically, the application allows the user to select the database from which the keywords will be extracted, the number of keywords of the taxonomy, the keyword field (authors keywords, index keywords etc.) and the child penalty factor.

After specifying these parameters, the keyword taxonomy will be created and visualized in the taxonomy viewer panel. This feature allows the user to interact with the taxonomy and modify its structure according to his own criteria. Using the panel, the user can modify any existing node by reattachment to a different parent node and deletion of an individual node or of entire subgraphs. Convenience functions such as previous action undo and exporting of taxonomies to different formats are also provided.

3.2   Technical Details

The Optimization Component. The algorithm described in Subsect. 2.2 is first used to create a matrix of similarities from which a measure of centrality can be derived for each node. Then, the taxonomy is grown by adding keyword nodes in the order of decreasing centrality using Eq. (2). In addition, as mentioned previously, the attachment of a node to a parent node is based on a weighted average of the similarities to the parent and higher ancestors of that node. Since the semi-automatic process allows the user to move or delete nodes, new similarity measures must be calculated after each modification as the structure and content of the taxonomy would also affect the final attachment strength (T in Eq. (2) may be different). Therefore, these nodes will often be attached to different parents. For example, imagine that the user moves a keyword node which has two child nodes and attaches it to a different parent. In this case, new similarities may need to be taken into account for the children of the removed node if the taxonomy contains different keywords from those it contained when these children were attached for the first time. In addition, the similarities between each of the child nodes and the candidate parent would also have changed since the path from their parent node to the root also changes and other nodes may now be more "similar" than the current parent node. This is true also in the case of the deletion of a node by the user. If this node is the parent of one or more nodes, the taxonomy generation process would have to be repeated for each of the child nodes.

Implementation. Our application is based on a client/server architecture which is similar to many existing web applications. Almost all computations are conducted on the server side which means that end users would not require advanced computing equipment to use this system. The server was implemented using the Python programming language with a relational database backend (the SQLite database was used for this system but other SQL database can be substituted). The frontend was implemented using Javascript, CSS and HTML and incorporated many modern frameworks such as Bootstrap [14] and the D3 library [2] which was used for visualizing the keyword taxonomy. Since our application contained a lot of Python and Javascript, Pico, a lightweight Remote Procedure Call (RPC) [22] library was used to make Python modules accessible from Javascript. A short demonstration video of the application can be found at: http://www.dnagroup.org/research.html.

4   Evaluation

This section describes the evaluation methodology used to compare the findings of the semi-automatic process with those of the fully automatic for generation of keywords taxonomies. The resulting comparisons are too many to be able to fit into the current format. However, an illustrative example concentrating the majority of the characteristics found in the evaluation process is presented.

4.1   Methodology

To properly evaluate our system, numerous consultations were conducted with domain experts as this is one of the most informative ways of evaluating an interactive data visualization system [13]. The evaluation methodology used to compare the fully-automated with the semi-automated taxonomies is as follows:

Step 1: The Scopus database was used to create a local database containing a large number keywords along with 119,393 document abstracts relevant to the renewable energy domain.

Step 2: The keywords retrieved were used by the tool to construct a taxonomy as described in Sect. 2.2 and the growth scores as shown in Eq. (5).

Step 3: Although, the application allows the creation of taxonomies with up to 1000 keywords, we generated taxonomies of 500 keywords in size to keep the discussions more manageable.

Step 4: First, taxonomies composed of 500 keywords were created using the **fully** automated taxonomy creation process. A default value of $cp = 0.5$ was used.

Step 5: A domain expert was asked to discuss the results of this approach and to identify all subtrees which in their view had weak face validity.

Step 6: The final step of the evaluation process was the creation of keyword taxonomies by the domain expert using the semi-automatic approach that our tool supports. The expert was allowed to change the cp value and move or delete nodes in order to produce a taxonomy which offers accurate visualizations of the dependencies between the technological fields.

4.2 Fully Automated Process

Figure 3 depicts a fully automated taxonomy for the node "electric power system". The resulting visualization contains a number of irregularities. These mainly appear in the branch starting from the "applied (co)" node. Here, it is difficult to explain what technological relationships this node and its child nodes ("probability distributions" and "probability") have with the general context of the "electric power system" field. In particular, it is even harder to explain why the node "applied (co)" is present at all, since there is no known technology or research domain with this name.

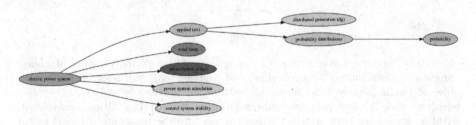

Fig. 3. Subtree for node "electric power system" produced using the fully automatic method

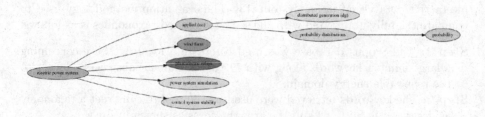

(a) Subtree "electric power system" using $cp=0.6$. The taxonomy remains the same with that produced using $cp=0.5$.

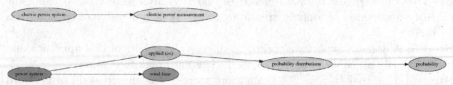

(b) Subtree "electric power system" using $cp=0.4$. The visualization obtained using this value of cp is not very rich. All the irrelevant nodes have been attached to another similar technology called "power system". Essentially, the problem has moved to another subtree of the broader taxonomy.

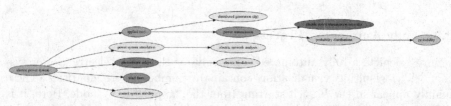

(c) Subtree "electric power system" using $cp=0.7$. This taxonomy is very similar to that obtained using $cp=0.5$. However, it is more informative because of the attachment of new related technologies (e.g. "power transmission", "electric network analysis"). It still contains irrelevant terms (e.g "applied(co)").

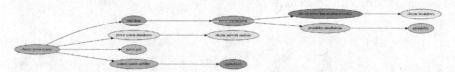

(d) Subtree "electric power system" using $cp=0.8$. Although the presence of nodes "probability distributions", "probability" and "applied (co)" still is questionable, the order of the taxonomic links seems better than all previous efforts (e.g. the "applied(co)" node is a leaf node now and can be easily deleted). In addition, new related scientific areas have been attached in comparison with the taxonomy obtained using $cp=0.5$ (e.g. "power grid", "electric breakdown").

Fig. 4. Taxonomies obtained for the "electric power system" subtree using the semi automatic method and different values of child penalty

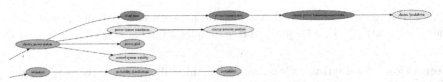

(a) Subtree "electric power system" after manual deletion of node "applied (co)" and reattachment of node "probability distributions" to node "estimation" belonging to another subtree using cp=0.8. The nodes "probability distributions" and "probability" still being connected to each other proving that the optimization function works properly.

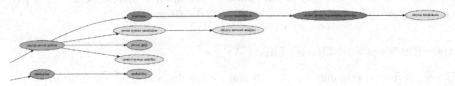

(b) Subtree "electric power system" using cp=0.8. The nodes "applied (co)" and "probability distributions" have been manually deleted. The child of the last ("probability" node) has been automatically attached to the technologically relevant node "estimation" demonstrating the validity of our optimization approach.

Fig. 5. Taxonomies obtained for the "electric power system" subtree using the semi automatic method and interaction

4.3   Semi Automated Process

Figures 4 and 5 depict taxonomies obtained for the "electric power system" subtree using the semi automatic method.

Child Penalty Adjustment: The visualization for subtree "electric power system" remains the same when the cp factor is equal to 0.6 (Fig. 4a). Nevertheless, when the value of $cp = 0.4$ is used the generated taxonomy is disappointing (Fig. 4b). Increasing the cp to 0.7 (Fig. 4c) offers a more informative visualization than that obtained using $cp = 0.5$. However, inconsistencies between the nodes of the taxonomy still exist with the presence of irrelevant terms. Setting the cp factor to 0.8 (Fig. 4d) yields the best taxonomy visualization among all others produced using the semi automatic approach with different values of cp. Using this setting the order of the taxonomic links seem to be more reasonable. In particular, the irrelevant "applied (co)" topic becomes a leaf node and deleting it will not affect the general structure of the subtree. Moreover, the "probability distributions" and "probability" nodes are now at the end of the path starting from the "wind farm" node and thus, they can be deleted or reattached to another subtree without causing any further changes to the relationships between the other nodes of the taxonomy. Another advantage of the visualization produced using $cp = 0.8$ is the presence of many related with electric power system topic technologies like electric breakdown, electric network analysis and power grid. These newly attached nodes make the taxonomy more explanatory.

Interaction: An improved visualization for this subtree can be obtained using $cp = 0.8$ in two ways: either by deleting the "applied (co)" node and reattaching the branch containing "probability distributions" and "probability" nodes to another more relevant subtree (Fig. 5a) or by deleting the "applied (co)" and "probability distributions" nodes (Fig. 5b). The second approach also results in an accurate taxonomy visualization since the optimization process attaches the "probability" node (a child of the deleted "probability distributions" node) to another more relevant scientific domain called "estimation".

4.4   Findings and Discussions

Based on the experiments conducted and the results obtained, some observations about the semi-automatic approach are:

(a) There are significant changes between the taxonomies obtained using different values of child penalty. In most cases child penalty values in the range of [0.4, 0.6] produce better results.

(b) There are some rare cases where values out of the range of [0.4, 0.6] generate the most accurate taxonomies. This can be observed at subtree "electric power system" (Fig. 3) where value of cp equal to 0.8 was able to produce the most informative visualization among all other values of cp that were explored for this subtree.

(c) Irregularities are often associated with leaf nodes in the taxonomies. This observation is quite reasonable if we consider that these nodes have low confidence scores and are either the less well known terms or the most generic. Reattaching the corresponding nodes to other subtrees using different values of child penalty or manually deleting them are two efficient and effective alternatives for generating accurate visualizations.

(d) The optimization process succeeds at attaching the children of the optimized node to related subtrees or retaining their connection in case that a technological dependency exists between them. Subtree "electric power system" (Fig. 5a) demonstrates this since after the manual reattachment of node "probability distribution" to node estimation its child and obviously strongly dependent node "probability" remains attached to it. In addition, the deletion of node "probability distributions" (Fig. 5b) causes the reattachment of node "probability" to node "estimation" similarly to the manual reattachment process. Clearly, there is strong connection between these two fields since one of the methods generally considered in estimation theory is the probabilistic approach [10].

(e) In some cases, manual deletion of the invalid node(s) is the only alternative for overcoming certain irregularities. Two specific cases were encountered: (i) where the subtree contains terms that do not represent a valid (existing or known) technological domain (ii) two or more keywords which seem to represent the same field or domain. In former is typified by the subtree attached to the "electric power system" node (Fig. 5a), where the term "applied (co)" had to be deleted. Experimenting with various values of cp for these subtrees proved incapable of producing accurate visualizations since either the problematic nodes could not be detached from the taxonomy or they were reattached to other subtrees with subsequent adverse side-effects.

(f) The accuracy of the results varied significantly between domains and between subtrees within the same domain. Hence, multiple actions (deletions, removals, optimizations, etc.) may be needed in the same subtree, which made the "undo" and "download" features very valuable for keeping track of all changes. For example, changing cp from 0.7 to 0.8 resulted in radical changes in the "electric power system" subtree (Fig. 4c and d respectively). Even the manual deletion of nodes "applied (co)" and "probability distributions" produced notable changes in the taxonomy. Saving the intermediate taxonomies or undoing proved very helpful in these circumstances.

5   Conclusion and Future Work

Although the fully automated taxonomy generation process proposed in [24] produced interesting results, it was still quite unstable and would occasionally yield unconvincing or inconsistent results. The web-based framework described in this paper represent an instantiation of a semi-automatic process that would allow valuable input from end-users to be incorporated into the technology visualization process. Our findings reveal that this technique allows the advantages of both approaches to be enjoyed. On the one hand, we benefit from the advantages of the automatic approach, namely the ability to quickly integrate the latest developments as well as to efficiently use large amounts of data; on the other hand, the semi-automatic approach allows the end-user fine-tune the automatically generated taxonomies which will help to remove noise and inconsistencies as well as support the incorporation of domain knowledge into the taxonomy generation process.

Future work will focus on improving the usability of the web-based platform by adding more advanced parameters for customization of the taxonomy generation process. In addition, the algorithm associated with the taxonomy creation as well as the enhancement of the taxonomy with suitable early growth indicators are the subjects of intense further investigations that we are currently pursuing.

References

1. Batagelj, V., Mrvar, A.: Pajek — analysis and visualization of large networks. In: Mutzel, P., Jünger, M., Leipert, S. (eds.) GD 2001. LNCS, vol. 2265, pp. 477–478. Springer, Heidelberg (2002). doi:10.1007/3-540-45848-4_54
2. Bostock, M., Ogievetsky, V., Heer, J.: D^3 data-driven documents. IEEE Trans. Vis. Comput. Graph. **17**(12), 2301–2309 (2011)

3. Chau, D.H., Kittur, A., Hong, J.I., Faloutsos, C.: Apolo: interactive large graph sensemaking by combining machine learning and visualization. In: KDD 2011, pp. 739–742 (2011)
4. Cilibrasi, R., Vitanyi, P.: The google similarity distance. IEEE Trans. Knowl. Data Eng. **19**(3), 370–383 (2007). doi:10.1109/TKDE.2007.48
5. Eades, P., Huang, M.L.: Navigating clustered graphs using force-directed methods. J. Graph Algorithms Appl. **4**(3), 157–181 (2000). doi:10.7155/jgaa.00029
6. Ellson, J., Gansner, E.R., Koutsofios, E., North, S.C., Woodhull, G.: Graphviz and dynagraph — static and dynamic graph drawing tools. In: Jünger, M., Mutzel, P. (eds.) Graph Drawing Software. Mathematics and Visualization, pp. 127–148. Springer, Heidelberg (2004). doi:10.1007/978-3-642-18638-7_6
7. Gretarsson, B., Bostandjiev, S., O'Donovan, J., Höllerer, T.: WiGis: a framework for scalable web-based interactive graph visualizations. In: Eppstein, D., Gansner, E.R. (eds.) GD 2009. LNCS, vol. 5849, pp. 119–134. Springer, Heidelberg (2010). doi:10.1007/978-3-642-11805-0_13
8. Herman, I., Melançon, G., Marshall, M.S.: Graph visualization and navigation in information visualization: a survey. IEEE Trans. Vis. Comput. Graph. **6**(1), 24–43 (2000). doi:10.1109/2945.841119
9. Heymann, P., Garcia-Molina, H.: Collaborative creation of communal hierarchical taxonomies in social tagging systems. Technical report 2006–10, Stanford InfoLab (2006). http://ilpubs.stanford.edu:8090/775/
10. Lehmann, E.L., Casella, G.: Theory of Point Estimation, vol. 31. Springer Science & Business Media, New York (1998)
11. Martino, J.P.: A review of selected recent advances in technological forecasting. Technol. Forecast. Soc. Change **70**(8), 719–733 (2003)
12. Milanez, D.H., de Faria, L.I.L., do Amaral, R.M., Leiva, D.R., Gregolin, J.A.R.: Patents in nanotechnology: an analysis using macro-indicators and forecasting curves. Scientometrics **101**(2), 1097–1112 (2014)
13. Oglic, D., Paurat, D., Gärtner, T.: Interactive knowledge-based kernel PCA. In: Calders, T., Esposito, F., Hüllermeier, E., Meo, R. (eds.) ECML PKDD 2014. LNCS (LNAI), vol. 8725, pp. 501–516. Springer, Heidelberg (2014). doi:10.1007/978-3-662-44851-9_32
14. Otto, M.: Bootstrap. http://twitter.github.io/bootstrap/. Accessed 30 May 2014
15. Porter, A.L.: Technology foresight: types and methods. Int. J. Foresight Innov. Policy **6**(1), 36–45 (2010). doi:10.1504/IJFIP.2010.032664. http://inderscience.metapress.com/content/M27482668H2V561L
16. Schmitt, N., Niepert, M., Stuckenschmidt, H.: BRAMBLE: a web-based framework for interactive RDF-graph visualisation. In: ISWC Posters & Demos 2010, p. 1 (2010)
17. Shannon, P., Markiel, A., Ozier, O., Baliga, N.S., Wang, J.T., Ramage, D., Amin, N., Schwikowski, B., Ideker, T.: Osgi alliance cytoscape: a software environment for integrated models of biomolecular interaction networks. Genome Res. **13**(11), 2498–2504 (2003)
18. Tom Sawyer Software: Tom sawyer visualization (2009)
19. Takeda, Y., Kajikawa, Y.: Optics: a bibliometric approach to detect emerging research domains and intellectual bases. Scientometrics **78**(3), 543–558 (2009). doi:10.1007/s11192-007-2012-5
20. Touchgraph: Touchgraph navigator, proprietary online application. http://www.touchgraph.com/navigator/

21. Tseng, Y.H., Lin, Y.I., Lee, Y.Y., Hung, W.C., Lee, C.H.: A comparison of methods for detecting hot topics. Scientometrics **81**(1), 73–90 (2009). doi:10.1007/s11192-009-1885-x

22. Walsh, F.: Pico. https://github.com/fergalwalsh/pico. Accessed 13 June 2014

23. Woon, W.L., Aung, Z., Madnick, S.: Forecasting and visualization of renewable energy technologies using keyword taxonomies. In: Woon, W.L., Aung, Z., Madnick, S. (eds.) DARE 2014. LNCS (LNAI), vol. 8817, pp. 122–136. Springer, Heidelberg (2014). doi:10.1007/978-3-319-13290-7_10

24. Woon, W.L., Henschel, A., Madnick, S.: A framework for technology forecasting and vizualization. In: International Conference on Innovations in Information Technology, IIT 2009, pp. 155–159. IEEE (2009)

25. Woon, W.L., Madnick, S.: Asymmetric information distances for automated taxonomy construction. Knowl. Inf. Syst. **21**(1), 91–111 (2009). doi:10.1007/s10115-009-0203-5

26. Woon, W.L., Zeineldin, H., Madnick, S.: Bibliometric analysis of distributed generation. Technol. Forecast. Soc. Change **78**(3), 408–420 (2011)

27. Yeo, W., Kim, S., Coh, B.Y., Kang, J.: A quantitative approach to recommend promising technologies for SME innovation: a case study on knowledge arbitrage from LCD to solar cell. Scientometrics **96**(2), 589–604 (2013). doi:10.1007/s11192-012-0935-y

28. Chen, Y.H., Chen, C.-Y., Lee, S.C.: Technology forecasting of new clean energy: the example of hydrogen energy and fuel cell. Afr. J. Bus. Manag. **4**(7), 1372–1380 (2010)

Decomposition of Aggregate Electricity Demand into the Seasonal-Thermal Components for Demand-Side Management Applications in "Smart Grids"

Andreas Paisios and Sasa Djokic[(✉)]

The University of Edinburgh, Edinburgh, Scotland, UK
a.paisios@sms.ed.ac.uk, sasa.djokic@ed.ac.uk

Abstract. Aggregate active and reactive power demands measured at 84 Scottish medium-voltage (MV) buses are used in this paper for the correlation and regression analysis, aimed at demand profiling and load decomposition. Demand profiles are presented with respect to the long-term seasonal variations, medium-term weekly and short-term diurnal cycles, allowing for the characterisation and presentation of load behaviour at different time-scales. The linear relationships between active and reactive power demands, temperature and power factor variations are quantified using regression analysis, on a per-hour of the day basis, as well as using a sliding-window regression approach for estimating relative coefficients within a seasonal moving window. The paper presents three different approaches for the decomposition of aggregate network demand into the temperature-dependent loads (i.e. thermal heating and cooling loads) and temperature-independent loads, providing important basic information for the application of the "smart grid" functionalities, such as demand-side management, or balancing of variable energy flows from renewable generation.

Keywords: Load decomposition and profiling · Smart grids · Demand-side management · Temperature-demand dependencies · Correlation and regression analysis · Sliding-window data analysis · Power-factor analysis

1 Introduction

Future electricity supply systems (so called "smart grids") will see significant increase of renewable-based distributed generation (DG), radical transformation of transmission and distribution networks and introduction of new highly efficient, intelligent and automated control, monitoring and communication infrastructures. This will be necessary to reduce CO_2 emissions and other drivers of climate change, while simultaneously maintaining high levels of security, sustainability and affordability of electricity supply. It is, however, widely recognised by all stakeholders that supply-side solutions alone will not be sufficient for the realisation of these challenging tasks. Additional strong support and contributions are expected to come from demand-side actions and measures, i.e. from a modified behaviour and physical demand for electricity through the consumer choice and evolved end-use of demand-manageable load

W.L. Woon et al. (Eds.): DARE 2016, LNAI 10097, pp. 116–135, 2017.
DOI: 10.1007/978-3-319-50947-1_11

and micro-generation. This will open major opportunities for a more direct and proactive system support, but will also result in profound changes in levels and nature of system-user interactions, shifting the actual system operating and loading conditions well outside the traditionally assumed ranges, limits and physical boundaries.

In spite of all anticipated changes, however, the simple fact is that a reliable and accurate assessment of the operation of both existing electricity networks and future "smart grids" cannot be performed if accurate models of aggregate system loads are not available. This is particularly true for the analysis of the effects and possible benefits of applying various demand-side management actions and schemes.

Demand-side management (DSM) is generally denoted as a set of measures, actions and interventions, initiated deliberately and with a specific purpose by end-users or network operators, or a third party (e.g. energy suppliers), aimed at changing, restructuring or rescheduling power demands of a group of loads, load sector(s), part of a system, or a whole system, in order to produce desired changes in the actual amounts and time patterns of power demands supplied at the dedicated point(s) of delivery for end-use consumption of electricity. Accordingly, before the specific DSM action or scheme should be applied, it is important to as accurately as possible identify the participation of the load targeted by the DSM in the total system demand and, equally, to estimate the potential effects of directly controlled DSM portion of the load on the improvement (or deterioration) of network performance.

Aggregate network demands are typically measured at primary distribution sub-stations, representing bulk load grid supply points (GSPs) connected at the medium-voltage (MV). Identification of the demand-manageable portion of the load in the total demand, therefore, requires decomposition and profiling of the GSP-measured aggregate demands. This paper builds on the initial results from [1] and presents further analysis of the temperature dependency of the total network demands, obtained as the sum of all measured individual GSP demands. In total, three different decomposition approaches are presented and discussed, all with low computational requirement and requiring only a limited set of input data. The presented analysis includes several metrics and indicators for quantifying temperature-demand dependencies, determining the portion of the load that is available for thermal heating and cooling purposes and identifying primarily resistive loads in the aggregate demand envelopes. These results are used as a basis for identifying contributions of the heating load to the total electricity demand during the winter period, as well as the participation of the cooling load in the total electricity demand during the summer period.

1.1 Overview of Previous Work

There is an increasing amount of work and effort dedicated to the analysis of electricity demands, where particularly the relationships between the demands and various weather/meteorological conditions, as well as human behaviour and economic factors have been analysed. Ultimately, the results of this work should be utilised by electricity industry, or transmission and distribution network operators for improved system operation and, in the recent years, as a means of devising and implementing optimal DSM schemes, which are an important part of "smart grid" functionalities.

Parameters that affect short-, medium- and long-term variations of electricity demand have been studied by a number of authors and include: the season of the year, the day of the week, the hour of the day, temperature, wind speed, relative humidity, public holidays, holiday seasons and geographic locations, as well as the factors relevant to a long-term analysis, such as economic development, climate change, technological innovations, population growth, urbanization, etc., [2–5].

Previous research included analysis of thermal electricity demands (e.g., for space and water heating loads), where contribution of thermal loads within a characteristic day/season of the year and for the characteristic hours of the day have been identified in [4, 6]. These characteristic hours have been identified as the points of changes in the daily power demands, either due to the group on/off switching of heating appliances, or as a consequence of habitual and socio-behavioural factors (e.g., differences in weekday-weekend working schedules and corresponding daily working hours).

Demand-side participation is an area which is recently receiving perhaps the most of the attention in the field of electricity demand analysis and optimization. For the implementation of various DSM actions and schemes, load identification in the residential, commercial and industrial sectors is of particular importance, and is typically aimed at a detailed profiling of variations of various load types (e.g., heating loads, power electronic loads, lighting loads, etc.) and their associations with weather and socio-economic factors. The examples include development of thermostatically controlled load models, or frequency controlled system reserve loads, [7].

Similarly, there are a number of models developed for electricity demand forecasting, including time series analysis, linear, nonlinear and multiple regression and correlation analysis, artificial intelligence and neural networks, grey-based approaches, expert systems and others, [8–12]. Long-term forecasting is important for system development/upgrading and planning of transmission and distribution networks.

Concerning load aggregation and disaggregation (decomposition), much of the previous work has been concentrated on either bottom-up approaches, or low-voltage end-user disaggregation approaches, such as non-intrusive load monitoring (NILM). The low-voltage disaggregation methods rely primarily on existing knowledge of the "signatures" of particular appliances and the choice sampling frequencies, employing methods such as pattern recognition and artificial neural networks (ANN), probabilistic approaches and others, [13, 14]. Aggregation methodologies, such as bottom-up load models, make use of empirical data collected from individual metering devices (both intrusive and NILM), as well as from survey-based socio-economic data and prior knowledge of appliance specifications in order to develop aggregate models starting from individual appliance usage, occupancy levels and user-specific schedules and behavioral information, [15, 16]. In this context, load models for space heating and cooling loads, water heating loads [6], lighting loads [17] and others have been developed and are used for determining the relative contributions from specific load categories, both at the low-voltage level, [18] and by aggregating these contributions at the medium-voltage level, [19]. Attempts for the development of load disaggregation methodologies at bulk supply points have also been made as in [20, 21], employing artificial neural networks (ANN), survey based analysis and weather related statistical analysis.

Building on the previous work discussed in [1, 22], this paper analyses three different approaches for the decomposition of electricity demands measured at MV buses in Scotland, UK. In all three cases, the main aim is to provide a more detailed analysis of the dependency of electricity demand on temperature variations from different temporal perspectives (seasonal, daily and hourly). This allows for the distinction between the dependencies at different time scales, which, in turn, will enable to study the corresponding effects in the context of short/medium/long term variations of electricity demand and related DSM applications in "smart grids".

2 Description of Available Measurement Datasets

Aggregate active power demands (in MW) and aggregate reactive power demands (in MVAr) were monitored at 84 Scottish medium voltage (MV) GSPs for a period of one year, from 01/04/2009 to 31/03/2010, with a half-hourly resolution (30-minute intervals). Temperature measurements (in °C) are taken from the University of Edinburgh's weather station.

2.1 Seasonal Variations

Measured active and reactive powers are presented as normalised values:

$$X'_i = \frac{X_i}{X_{max}} \tag{1}$$

where: X'_i is the i-th normalised value, X_i is the actually measured i-th value and X_{max} is the maximum value of each dataset (i.e. maximum active/reactive power recorded in the considered period). Power factor values are calculated using (2):

$$pf_i = \frac{P_i}{\sqrt{P_i^2 + Q_i^2}} \tag{2}$$

where: pf_i is the i-th power factor value, P_i and Q_i are the actual measured i-th values of active and reactive powers. Temperature values are converted from degrees Celsius (°C) to Kelvin (K), in order to prevent the occurrence of negative values in the dataset. Figures 1a–d show the seasonal variations of active power, reactive power, power factor and temperature, respectively. The daily mean, daily maximum and daily minimum values of each variable are also shown, as well as the mean daily smoothed values, which are calculated using a moving-average sliding window filter, denoted as "sliding window" (SW) values and calculated using (3):

$$V_{SW_d} = \frac{\sum_{i-14}^{i+14} \overline{V_d}}{29} \tag{3}$$

where: V_{SW_d} is the sliding window smoothed value at day-d and $\overline{V_d}$ is the mean daily value at day-d. The window length is set at ±14 days, i.e. ±2 weeks.

Figure 1 shows that the four variables exhibit higher frequency short-term variations, which are imposed on the lower frequency long-term variations. These long-term seasonality is more evident when considering the SW-values, particularly for active power demand and temperature and, to a lesser extent, for power factor values.

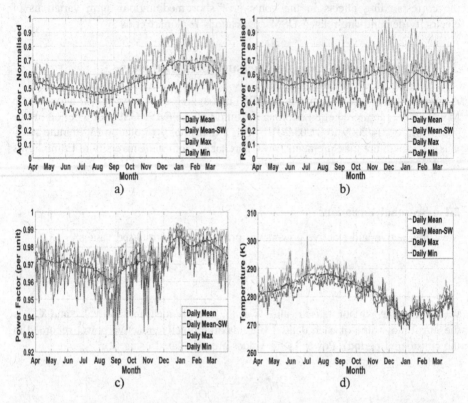

Fig. 1. Seasonal variations of total network active (a) and reactive (b) power demands (sums of all 84 GSP measurements), corresponding power factor (c) and temperature (d) values, without distinction between the weekdays and weekends

As the measured total system demand corresponds to a cold-climate geographic area of Scotland, for which electricity demand of the thermal loads is predominantly determined by the resistive heating loads in winter (there is a smaller contribution of non-resistive cooling load during the summer), the low-frequency SW-values for temperature and active power demand are opposite in phase, indicating that the maximum network active power demand occurs around the minimum temperature and vice versa. The reactive power demand curves exhibit much bigger short-term variations and have almost no long-term periodical variations, which suggest that power factor values during the winter period increase due to higher relative contributions of resistive heating loads (assumed to have unity power factor). This is further discussed in Sect. 4.3. The short-term periodicity in active and reactive power demands corresponds to a seven-day weekly cycle, due to the differences in demands between weekdays and weekends. This is discussed in more detail in Sect. 2.2.

2.2   Diurnal Variations and Weekday/Weekend Load Profiles

Figures 2a–d illustrate diurnal variations of active power, reactive power, power factor and temperature, plotted as the average annual values for each half-hour of the day (30-minute resolution), for all days (365 days), weekdays only (261 days) and weekends only (104 days) in a calendar year.

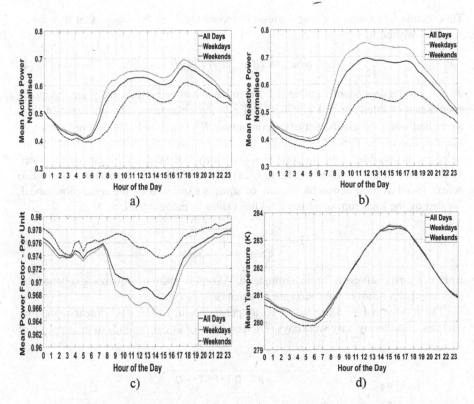

Fig. 2. Diurnal "perspective" of seasonal variations, represented as "average days", i.e. daily load curves for: (a) active power, (b) reactive power, (c) power factor, (d) temperature

The results presented in Fig. 2 demonstrate that there are clear differences in active power, reactive power and power factor values between weekdays and weekends. The differences can be attributed primarily to the underlying socio-behavioural factors (weekly working-hours schedule) and are therefore absent in the case of temperature (Fig. 1d). As an example, the morning increase of active power demand on weekends occurs later and has a lower gradient than on weekdays, while weekdays are characterised by a higher overall normalised demand, both for active and reactive powers.

As mentioned, this 7-day cycle can also be seen in Fig. 1 and is particularly pronounced for reactive power, for which there exists a higher deviation between normalised average demand for weekdays and weekends (Figs. 1b and 2b). Regarding the power factor, higher average values are found during weekends, with stronger differences for the period between 07:00 h to 18:00 h. This shows that although both active and reactive power demands are reduced on weekends, the reduction is more pronounced for reactive power than for active power.

3   Linear Regression Analysis

This section introduces a linear regression model used for the analysis in this paper, which is written in general form as:

$$y = a + x\beta + \varepsilon \approx f(x, \beta, a) \tag{4}$$

where: y is the dependent variable, x is the independent variable, a and β are the model coefficients (y-intercept and gradient) and ε is the error term. The goodness-of-fit is quantified using the coefficient of determination, R^2, which indicates the proportion of the variance of the dependent variable that can be attributed to the variability of the independent variable, or the proportion of variability in the dataset that can be explained by the model of the least squares fit, [23, 24]. The coefficient of determination can also be expressed as the fraction of the sum of squares explained by the regression model, divided by the total sum of squares, taking values in range from 0 to 1:

$$R^2 = \frac{SSreg}{SStot} = 1 - \frac{SSerr}{SStot} \tag{5}$$

where: R^2 is coefficient of determination, $SSreg$ is sum of squares explained by regression and $SStot$ is the total sum of squares.

The results in Fig. 3 are based on a "per-half-hour of the day" linear regression analysis, considering only weekdays. The goodness of fit corresponds to the strength of

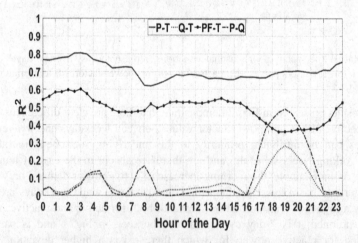

Fig. 3. Linear regression results: R^2 values for active power and temperature (P-T), reactive power and temperature(Q-T), power factor and temperature (PF-T) and active power and reactive power (P-Q), on a per-half-hour of the day basis, for weekdays only.

the seasonal correlations between the presented variables, at each half-hour of the day. Strong correlations are found between active power and temperature and, to a bit lesser extent, between power factor and temperature. Noticeable correlations between active and reactive power, as well as reactive power and temperature, are found only during the afternoon-evening period, between 16:00 and 22:00 h.

3.1   Sliding-Window Linear Regression for Selected Hours of the Day

Figures 4a–d show the resulting beta coefficients (β) and R^2 values for active power demand and temperature at two characteristics hours of the day (03:00 and 17:00), throughout the year, calculated using a sliding-window linear regression model and considering all days, i.e. a linear best fit (4) applied to each day of the year at a selected hour, using y and x values within a sliding window range of ±14 days (±2 weeks).

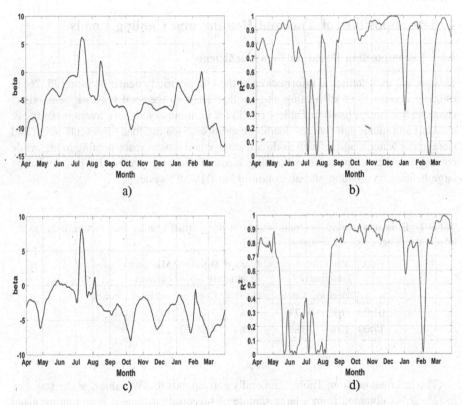

Fig. 4. Sliding-window (SW) linear regression analysis for two characteristic hours of the day: (a) beta at 03:00 h, (b) R^2 at 03:00 h, (c) beta at 17:00 h and (d) R^2 at 17:00 h

This analysis aims to capture the sensitivity of the correlations between active power demand and temperature, considering smaller samples of the datasets throughout the year, unlike the results presented in Fig. 3. It can be seen, for example, that the existing strong negative correlations at 03:00 h start to change at the beginning of July and oscillate between positive and negative gradients (beta values) until the beginning of September, indicating periods for which heating loads are significantly reduced and cooling loads may become comparable to, or higher than the heating loads. Reduced R^2 values and higher positive beta values at 03:00 and 17:00 coincide with the highest measured temperatures, indicating periods of minimum ("base") heating load. The results in Fig. 4 also show the yearly periods for which "saturation" of the demand-temperature dependencies are found (cold winter days during the period from end-January to mid-February), when maximum heating loading load is reached and heating demand does not increase with further decrease of temperature, or during hot summer days, e.g. July-August, when the gradients of best fits are close(r) to zero. Furthermore, the results indicate periods of 'atypical" demand changes, e.g. during the Easter and Christmas holiday periods. These results are used in Sect. 4.1 for the decomposition of the total demand into the heating and cooling loads.

4   Decomposition of Thermal Heating and Cooling Loads

4.1   Decomposition Based on Beta Coefficients

This section uses statistical information on the UK electricity demands from [25, 26] to estimate contributions of heating and cooling loads to the total network demands at characteristic hours, listed in Table 1 (in %) for minimum (summer), average (spring & autumn) and maximum (winter) loading conditions. Heating load represents direct and storage hot water loads, as well as direct, storage and top-up space heating loads, while cooling load represents air-conditioning loads of different sizes and types, including large heating, ventilation and air-conditioning (HVAC) systems.

Table 1. Estimated percentage contributions of the thermal heating and cooling loads to the total network active power demands, [25, 26].

Hour	Minimum (summer)		Average (spring & autumn)		Maximum (winter)	
	Heating	Cooling	Heating	Cooling	Heating	Cooling
03:00	6%	7%	18%	2%	29%	0%
17:00	10%	12%	15%	4%	23%	0%

The information from Table 1 generally corresponds to SW-values, as the statistics in [25, 26] is obtained from a large sample of surveyed customers, representing mean demands for the selected hours and loading conditions. This information and results from Sect. 3.1 are used to specify the following analytical relationships for the resulting heating and cooling loads over the course of the year:

$$P_{heat\_hr_i} = P_{base\_heat\_hr} + C_{heat\_hr} \cdot beta_{hr_i} \cdot (T_{hr_i} - T_{max\_hr}) \qquad (6a)$$

$$P_{cool\_hr_i} = P_{max\_cool\_hr} - C_{cool\_hr} \cdot beta_{hr} \cdot (T_{hr_i} - T_{max\_hr}) \qquad (6b)$$

where: $P_{heat\_hr_i}$ and $P_{cool\_hr_i}$ are heating and cooling demands at hour (hr) 03:00 and 17:00 on the i-th day, $P_{base\_heat\_hr}$ and $P_{max\_cool\_hr}$ are minimum (base) heating demand (2nd column in Table 1) and maximum cooling demand (3rd column in Table 1), C_{heat_hr} and C_{cool_hr} are weighting coefficients (for adjusting values calculated in Sect. 3.1 with values from Table 1, $C_{cool\_0300} = 0.75$, $C_{cool\_1700} = 0.7$, $C_{heat\_0300} = 0.6$, $C_{heat\_1700} = 0.4$, $beta_{hr_i}$ is the beta coefficient on the i-th day at the hour hr, T_{hr_i} is the i-th value of SW-temperature at hour hr, and T_{max_hr} is the maximum SW-smoothed temperature at hour hr.

The values of decomposed thermal cooling and heating demands at 03:00 h and 17:00 h calculated using (6), for the whole year and representing corresponding seasonal changes, are plotted in Fig. 5 as normalised values, i.e. as a ratio to the maximum/peak annual demand, and as a % of the total actual demand at that hour.

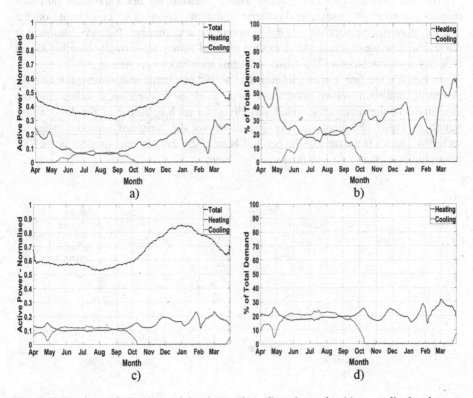

Fig. 5. SW-values of decomposed heating and cooling demands: (a) normalised values at 03:00 h, (b) as % of total actual demand at 03:00 h, (c) normalised values at 17:00 h and (d) as % of total actual demand at 17:00 h

4.2 Decomposition Based on the Seasonal Range of Active Power Demand Variations

The second approach for the decomposition of total active power demand into thermal heating loads is presented in this section. Figure 6a shows the normalised active power demand at 17:00 h, with the corresponding temperature measurements at the same hour, for all days of the year. Figure 6b shows the same data, but instead of the actual measured normalised demand and absolute temperature values, the sliding window (SW) smoothed values are presented, calculated using (3).

When considering the smoothed active power and temperature values, the annual datasets show a characteristic "loop", allowing to make distinction between active power demand as a function of temperature when considering the path from winter to summer (lower band) and from summer to winter (upper band), as shown in Fig. 6b. This is due to the phase differences in the seasonal periodicity of active power demand and temperature variations and the possible effects of solar irradiance levels and the presence of other seasonally variable active power loads in the system. The characteristic loops are analysed and discussed in more detail in [22].

The use of smoothed (SW) active power demand for decomposition purposes reduces the effect of stochastic short-term variations, allowing to concentrate on the general inherent characteristics of the seasonal demand changes. It can be shown that the residuals between actual and smoothed demand values are normally distributed and lack any autocorrelations [22], which confirms their stochastic nature.

In Fig. 6b, the two horizontal lines denote 95th percentile maximum value and 5th percentile minimum value, which are both used as the reference values for the assessment of the range of seasonal variations, for each half-hour of the day, as presented in Fig. 7. Furthermore, Fig. 7 also shows the weighted seasonal range of variations, which is the ratio of the per-half-hour of the day seasonal range over the 95th percentile maximum value at the same half-hour.

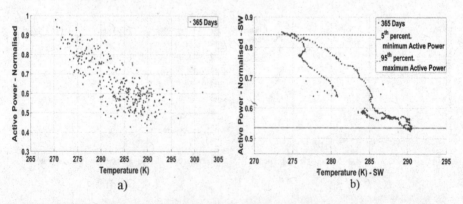

Fig. 6. Active power demand as a function of temperature at 17:00 h, (a) normalized active power demand, (b) SW-values of normalized demand and temperature with maximum and minimum percentile values

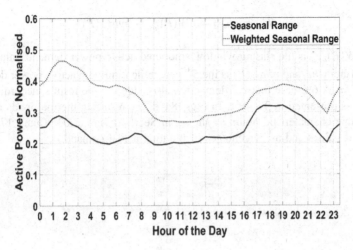

Fig. 7. Seasonal range of variations and weighted seasonal range of variations of active power demand per half-hour of the day

The corresponding equations for seasonal range and weighted seasonal range of active power demand variations are:

$$SR_{hr}(P) = max_{hr}(P) - min_{hr}(P) \tag{7}$$

$$WSR_{hr}(P) = \frac{max_{hr}(P) - min_{hr}(P)}{max_{hr}(P)} \tag{8}$$

where: hr is the hour of the day, P is the sliding window active power demand and max and min are the 95th percentile maximum value and 5th percentile minimum value, respectively. The seasonal range gives an indication of the seasonally variable portion of the total demand per half-hour of the day, which is assumed to be the part of the demand that accommodates changes in loading due to thermal heating and cooling purposes. The weighted seasonal range shows the same metric with respect to the total maximum active power demand, resulting in a higher peak during night hours (23:30–04:00), compared to the evening peak (16:00–22:00), clearly showing that although the demand during the night hours is lower, it has a higher proportion of variable, i.e. seasonal thermal load (representing so called "economy 7" customers in Scotland, with electrical heating load connected during the night hours).

It should be noted that in this section no distinction is made between heating and cooling load demands, which are presented together as the (total) thermal load demand. Accordingly, thermal demand for each half-hour of the day and day of the year is defined as the difference between the SW demand values and the 5th percentile minimum values, i.e. it is assumed that any increase of demand over this minimum value occurs due to connection of thermal loads. Effectively, this overestimates contribution of actual thermal loads:

$$Thermal_{hr,d} = SW(P)_{hr,d} - min_{hr}(P) \qquad (9)$$

where: $SW(P)_{hr,d}$ is the sliding-window smoothed active power demand value at day-d and half-hour-hr, and $min_{hr}(P)$ is the 5th percentile minimum active power demand at half-hour-hr. A 6% base thermal demand is also added to the results, according to the percentages presented in Table 1. In Figs. 8a–d, the resulting thermal loads are given for the same characteristic hours as shown in Sect. 4.1, i.e. at 03:00 and 17:00 h, as normalised values and as percentages of the total actual demand.

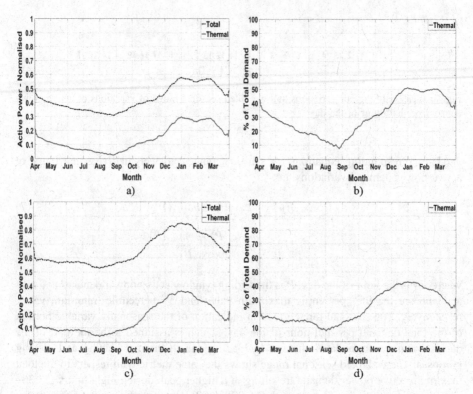

Fig. 8. SW-values of decomposed thermal demand: (a) normalised values at 03:00 h, (b) as % of total demand at 03:00 h, (c) normalised values at 17:00 h, (d) as % of total demand at 17:00 h

As shown in Fig. 8, this approach produces estimates of thermal demand that directly reflect the seasonality of the total demand, since the estimations are based on simply removing minimum (base) demand, calculated for each half-hour of the day.

Figure 9a shows the average, 95th percentile maximum and 5th percentile minimum values of the percentage of thermal demand to total demand for each hour of the day. Figure 9b shows the same statistics from a seasonal perspective, i.e. 95th percentile maximum and 5th percentile minimum values for the percentage of thermal demand, for each day of the year (average daily values).

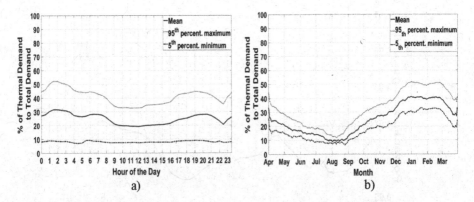

Fig. 9. Percentage of thermal demand to total demand, (a) diurnal "perspective": mean, maximum and minimum values per half-hour of the day, (b) seasonal "perspective": mean, maximum and minimum values per day of the year (average daily values)

4.3   Decomposition Based on Power Factor Analysis

The third approach for the decomposition of the total active power demand into thermal heating loads considers average minimum power factor values at each half-hour of the day, in an attempt to identify the contribution of the purely resistive loads (with unity power factor, e.g. space and water heating loads) to the variations of active power demand. Figure 10 shows the total system active and reactive power demands for one year at 17:00 h. The diagonal dotted-line corresponds to the average minimum power factor for the considered period, which is further illustrated in Fig. 11.

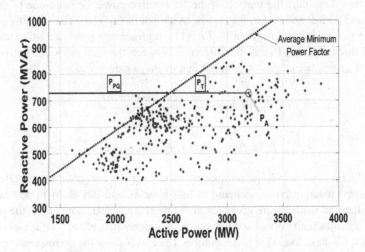

Fig. 10. Graphical representation of power-factor based decomposition approach. For each actual measured active power value (P_A), demand is separated into the purely resistive thermal loads (P_T) and active & reactive loads (P_{PQ}, determined by the average minimum power factor)

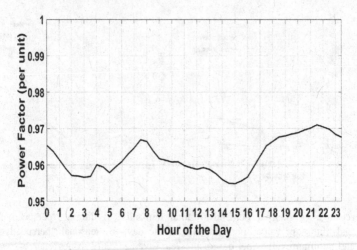

Fig. 11. Average minimum power factor values for each half-hour over the course of a calendar year

As shown in Fig. 10, non-normalised (i.e. absolute) values of active and reactive power demands are used in order to preserve the actual power factor values, calculated using (2). Again, SW non-normalised values with lower day-to-day fluctuations are used to estimate average minimum power factor values, in order to allow for direct comparison with the results from the previous two sections. Figure 11 shows the corresponding average minimum power factor values for each half-hour of the day.

Afterwards, the further decomposition is based on a simple assumption that for each measured reactive power demand, the reference (i.e. average minimum) power factor values determine the portion of the total active power demand due to the loads that demand both P and Q, i.e. P_{PQ} loads, as shown in Eq. (10). The remaining part of the active power demand is given by Eq. (11), representing purely resistive loads (i.e. electrical thermal loads, P_T) as the difference between the total measured active power demand (P_A) and active power demand due to P_{PQ} load.

$$P_{PQ}(d, hr) = \frac{pf_{min}(hr) \times Q(d, hr)}{\sqrt{1 - pf_{min}^2}} \tag{10}$$

$$P_T(d, hr) = P_A(hr, d) - P_{PQ}(d, hr) \tag{11}$$

where: $pf_{min}(hr)$ is the average minimum power factor at half-hour hr and $Q(d, hr)$ is the measured reactive power demand at half-hour hr and day d. In Figs. 12a–d, the resulting thermal loads P_T are given for the same characteristic hours as in the previous sections, again as both normalised values and as percentages of the total actual demand, with added 6% base load (Table 1). Figures 13a and b show the corresponding diurnal and seasonal percentage estimations, as in the previous sections.

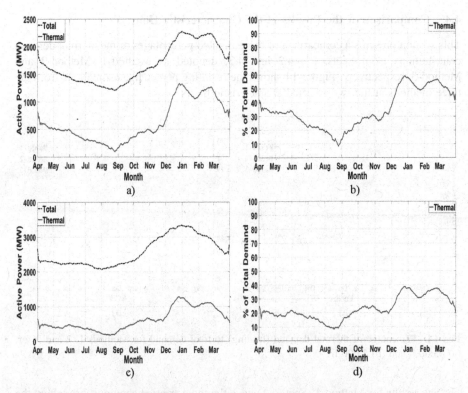

Fig. 12. SW-values of decomposed thermal demand: (a) normalised values at 03:00 h, (b) as % of total demand at 03:00 h, (c) normalised values at 17:00 h, (d) as % of total demand at 17:00 h

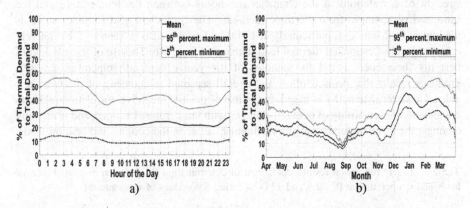

Fig. 13. Percentage of thermal demand to total demand, (a) diurnal "perspective": mean, maximum and minimum values per half-hour of the day, (b) seasonal "perspective": mean, maximum and minimum values per day of the year

4.4   Comparison of the Results at the Characteristic Hours

This section presents a comparison of the estimated percentages of the thermal demand contributions from Sects. 4.1, 4.2 and 4.3, denoted as Method-1, Method-2 and Method-3, respectively. Figure 14 shows the resulting percentage estimations from the three methods at the two selected characteristic hours.

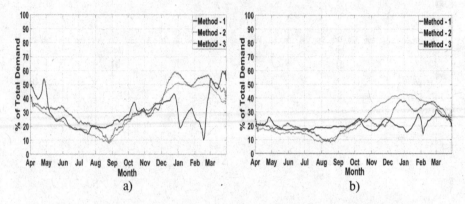

Fig. 14. Percentage of thermal demand (heating) to total demand for Methods 1, 2 and 3 for: (a) 03:00 h and (b) 17:00 h

The results for Method-1 (Sect. 4.1) have the most erratic behaviour throughout the year and generally fail to follow the yearly cycle related to the changes in load composition (and corresponding demands) due to the variations in temperature levels. This is also reflected in the linear regression results presented in Table 2, quantifying strength of correlations at the characteristic hours between the temperature and the estimated percentages from the three methods. The coefficient of determination R^2 is the lowest for Method-1, particularly at 17:00 h. This can also be seen by the relatively constant level of estimated thermal load demand in Fig. 14b. The use of the gradient of best fits (beta coefficients) for samples of the yearly demand-temperature datasets produces results that do not follow the overall seasonal relationship of the two variables. However, Method-1 is based on the use of positive/negative correlation and beta coefficient, i.e. sensitivity of demand changes with temperature changes, and is the only amongst the three methods that can distinguish between the cooling and heating loads.

Table 2. Linear regression results, coefficient of determination R^2, between estimated thermal loads and temperature at 03:00 h and 17:00 h (using SW-values of temperature)

	R^2 values	
	03:00 h	17:00 h
Method − 1	0.26	0.17
Method − 2	0.96	0.89
Method − 3	0.95	0.87

The results in Fig. 14 suggest that Method-1 should be made less sensitive to temperature variations, by e.g. selection of lower coefficients in Eq. (6), as the noticeable variations in the estimated values of heating load in the period from December 2009 to February 2010 were related to demand reduction around the Christmas holiday, followed by a colder than usual January-February 2010 (eighth coldest since 1910), resulting in the previously discussed saturation of heating load demands and inability of beta coefficient to correctly represent the underlying changes, or lack of changes, of thermal demands.

Method-2 and Method-3 are more successful in capturing the overall underlying seasonality of the changes in thermal demand, demonstrated by very high R^2 values for both methods and for both characteristic hours in Table 2. The estimated contributions of thermal loads to the aggregate demand by these two methods are closer for 03:00 h than for 17:00 h. This can be explained by the fact that Method-3 distinguishes between purely resistive loads and other loads in the system, while Method-2 allocates all changes in active power demand to thermal loads. The difference in estimations is higher for periods of the day when electricity is consumed at commercial and industrial establishments (still open for business) and also for residential use of appliances with reactive power requirements, such as electronic equipment, lighting, etc. When such loads are switched off (significant number of commercial industrial premises are closed during the night hours and most of the people are sleeping), resistive heating loads will have higher overall penetration in the system (particularly during the winter periods), and the two methods will then tend to produce similar estimations (December to April in Fig. 14a).

5 Conclusions

This paper presents three different and relatively simple methods for the decomposition of the total active power demands into the percentage contributions from non-thermal and thermal electrical loads, where latter are further separated into heating and cooling loads. While the presented results show certain differences between different methods, it should be noted that the paper was more focused on the analysis of the general characteristics and suitability of the considered methodologies, than on improving their accuracy. For example, the analysis in the paper demonstrates that power factor-related analysis can be used to distinguish between active power demands of purely resistive (heating) loads, and active power demands of loads that consume both active and reactive power. Nevertheless, the authors are working on the improvement of the accuracy of the presented methods.

Decomposition of the aggregate demands in different types of loads is required for the correct specification and implementation of a range of "smart grid" functionalities, where of particular importance are demand-side management (DSM) applications, aimed at controlled on/off switching and regulation of specific (target) load types, e.g. [27, 28]. The short- and longer-term variations and changes in the load structure and load composition present a challenge for the implementation of DSM schemes and this paper provides a contribution to this topic, through a detailed analysis of the dependencies and seasonal variations of heating and cooling demands, directly correlating temperature variations with the changes in active and reactive powers, i.e. different types of loads responsible for these changes. The paper also analyses these changes over different temporal scales (annual, seasonal, daily and hourly), allowing to study the effects of load and demand changes for the corresponding "smart grid" DSM applications, e.g. an hour-ahead, or a day-ahead scheduling of loads. In this context, the results presented in the paper are of particular importance for the implementation of DSM functionalities in residential and commercial load sectors, as they may provide (close to) real-time estimations of the expected heating and cooling loads in the total demands, based on the weather/temperature forecasts and expected socio-behavioural factors, such as weekdays, weekends, holidays, etc.

References

1. Paisios, A., Djokic, S.: Load decomposition and profiling for "smart grid" demand-side management applications. In: ECML PKDD 2013 Workshop on Data Analytics for Renewable Energy Integration (DARE 2013), Prague, Czech Republic (2013)
2. Hor, S.L., Watson, S., Majithia, S.: Analyzing the impact of weather variables on monthly electricity demand. IEEE Trans. Power Syst. **20**, 2078–2085 (2005)
3. Noureddine, A.H., Alouani, A.T., Chandrasekaran, A.: A new technique for short-term residential electric load forecasting including weather and lifestyle influences. In: Proceedings of the 35th Midwest Symposium on Circuits and Systems, UK, vol. 2, pp. 1419–1427 (1992)
4. GPG310 - Degree Days for Energy Management - A Practical Introduction, Government Action Energy Program, New Action Energy, Carbon Trust, London (2004)
5. Rahman, S., Hazim, O.: A generalized knowledge-based short-term load-forecasting technique. IEEE Trans. Power Syst. **8**, 508–514 (1993). IEEE Press
6. Lane, I.E., Beute, N.: A model of the domestic hot water load. IEEE Trans. Power Syst. **11**, 1850–1855 (1996)
7. Xu, Z., Ostergaard, J., Togeby, M., Marcus-Moller, C.: Design and modeling of thermo-statically controlled loads as frequency controlled reserve. In: Power Engineering Society General Meeting, pp. 1–6 (2007)
8. Park, D.C., El-Sharkawi, M.A., Marks, R.J., Atlas, L.E., Damborg, M.J.: Electric load forecasting using an artificial neural network. IEEE Trans. Power Syst. **6**(2), 442–449 (1991)
9. Rahman, S., Hazim, O.: A generalized knowledge-based short-term load-forecasting technique. IEEE Trans. Power Syst. **8**, 508–514 (1993). IEEE Press
10. Al-Hamadi, H.M., Soliman, S.A.: Long-term/mid-term electric load forecasting based on short-term correlation and annual growth. Electr. Power Syst. Res. **74**, 353–361 (2005)
11. Al-Alawi, S.M., Islam, S.M.: Principles of electricity demand forecasting - part I methodologies. Power Eng. J. **10**, 139–143 (1996)
12. Hyndman, R.J., Fan, S.: Density forecasting for long-term peak electricity demand. IEEE Trans. Power Syst. **25**, 1142–1153 (2010). IEEE Press
13. Laughman, C., Lee, K., Cox, R., Shaw, S., Leeb, S., Norford, L., Armstrong, P.: Power signature analysis. IEEE Power Energy Mag. **1**(2), 56–63 (2003)
14. Liang, J., Ng, S., Kendall, G., Cheng, J.: Load signature study-part I: basic concept, structure, and methodology. IEEE Trans. Power Deliv. **25**(2), 551–560 (2010)
15. Capasso, A., Grattieri, W., Lamedica, R., Prudenzi, A.: A bottom-up approach to residential load modeling. IEEE Trans. Power Syst. **9**(2), 957–64 (1994)

16. Paatero, J.V., Lund, P.D.: A model for generating household electricity load profiles. Int. J. Energy Res. **30**(5), 273–290 (2006)
17. Stokes, M., Rylatt, M., Lomas, K.: A simple model of domestic lighting demand. Energy Build. **36**(2), 103–116 (2004)
18. Collin, A.J., Tsagarakis, G., Kiprakis, A.E., McLaughlin, S.: Multi-scale electrical load modelling for demand-side management. In: 3rd IEEE PES Innovative Smart Grid Technologies Europe (ISGT Europe) (2012)
19. Collin, A.J., Hernando-Gil, I., Acosta, J.L., Djokic, S.Z.: An 11 KV steady state residential aggregate load model. Part 1: aggregation methodology. In: IEEE Trondheim PowerTech (2011)
20. Xu, Y., Milanovic, J.V.: Artificial-intelligence-based methodology for load disaggregation at bulk supply point. IEEE Trans. Power Syst. **30**(2), 795–803 (2014)
21. Hobby, J.D., Tucci, G.H.: Analysis of the residential, commercial and industrial electricity consumption. In: 2011 IEEE PES Innovative Smart Grid Technologies Asia (ISGT), Perth, WA, pp. 1–7 (2011)
22. Paisios, A., Ferguson, A., Djokic, S.: Solar analemma for assessing variations in electricity demands at MV buses. In: Med Power 2016 Conference (2016)
23. Buda, A., Jarynowski, A.: Life-time of Correlations and its Applications, vol. 1 (2010)
24. Steel, R.G.D., Torrie, J.H.: Principles and Procedures of Statistics, New York, pp. 187–287 (1960)
25. Energy Consumption in the UK, Department of Energy and Climate Change, UK Government, London (2012)
26. Pout, C., MacKenzie, F., Olloqui, E.: The impact of changing energy use patterns in buildings on peak electricity demand in the UK, building research establishment. Technical report 243 752 (2008)
27. Hayes, B.P., H.-Gil, I., Collin, A.J., Harrison, G., Djokic, S.Z.: Optimal power flow for maximizing network benefits from demand-side management. IEEE Trans. Power Syst. **29**(4), 1739–1747 (2014)
28. Djokic, S.Z., Papic, I.: Smart grid implementation of demand side management and micro-generation. Int. J. Energy Optim. Eng. **1**(2), 1–19 (2012)

Author Index

Al Junaibi, Reem 22
Alibasic, Armin 22
Álvarez, Mauricio A. 43
Ang, Wee Horng 67
Aung, Zeyar 22

Catalina, Alejandro 31

Deng, Vicki 67
Djokic, Sasa 116
Donker, Hilko 54
Dorronsoro, José R. 31

Gräfe, Gunter 54

Heinemann, Detlev 78
Heinermann, Justin 92
Hug, Gabriela 10

Jalali, Mohammad S. 67

Karakatsanis, Ioannis 102
Koch, Stephan 10
Kogler, Alexander 1
Kramer, Oliver 78, 92

Lässig, Jörg 92
Lee, Yang 67
Lehner, Wolfgang 54

Madnick, Stuart 67
Mistree, Dinsha 67

Omar, Mohammad Atif 22

Paisios, Andreas 116

Siegel, Michael 67
Strong, Diane 67

Thoß, Anna 54
Torres-Barrán, Alberto 31
Traxler, Patrick 1
Tsoupos, Alexandros 102

Ulbig, Andreas 10
Ulbricht, Robert 54

Wang, Richard 67
Wolff, Björn 78
Woon, Wei Lee 22, 102

Zufferey, Thierry 10
Zuluaga, Carlos D. 43

Printed in the United States
By Bookmasters